NO BONE UNTURNED

NO BONE UNTURNED

The Adventures of a Top
Smithsonian Forensic Scientist
and the Legal Battle for
America's Oldest Skeletons

JEFF BENEDICT

HarperCollins*Publishers*

NO BONE UNTURNED. Copyright © 2003 by Jeff Benedict. All rights reserved. Printed in the United States of America. No part of this book may be used or reproduced in any manner whatsoever without written permission except in the case of brief quotations embodied in critical articles and reviews. For information, address HarperCollins Publishers Inc., 10 East 53rd Street, New York, NY 10022.

HarperCollins books may be purchased for educational, business, or sales promotional use. For information, please write: Special Markets Department, HarperCollins Publishers Inc., 10 East 53rd Street, New York, NY 10022.

FIRST EDITION

Designed by Nancy B. Field

Printed on acid-free paper

Library of Congress Cataloging-in-Publication Data is available upon request.

ISBN 0-06-019923-7

03 04 05 06 07 ❖/RRD 10 9 8 7 6 5 4 3 2 1

To Lydia,
The one I run to in the dark for light,
The one whose faith gives me flight,
The one who takes away my fears,
The one who wipes away our children's tears,
The one whose strength knows no end,
The one who is my Juliet and my best friend,
The one.

CONTENTS

AUTHOR'S NOTE

The book you are about to read is not the one I set out to write when I began my research in the summer of 1999. At that point I was planning a book on the landmark lawsuit *Robson Bonnichsen et al* v. *U. S. et al* that had been filed by a group of scientists against the federal government. The suit arose after one of America's oldest complete human skeletons—the 9,800-year-old Kennewick Man—surfaced in 1996 near the Washington-Oregon border. Scientists from the Smithsonian Institution wanted to study the remains. But the federal government seized the skeleton and prohibited study.

After reading up on the background of the case, I traveled to Portland, Oregon, to conduct research at the law firm that was representing the scientists. I spent hours interviewing attorneys and poring over stacks of legal documents. At the end of my visit, one lawyer said, in passing, "Sometime you ought to get to know some of our plaintiffs. They are really interesting people."

I asked what was so interesting about the plaintiffs. The attorney talked almost exclusively about Dr. Douglas Owsley, the Smithsonian scientist who was the driving force behind the lawsuit. The lawyer said that Owsley was probably the world's top expert on human skeletons. The federal government relied on him like a special agent. The FBI had sent him to Waco after the Branch Davidians died in the fire. The State Department dispatched him to the Balkans after the Bosnian War. The U.S. embassy in Guatemala called him in after two American journalists were murdered there. And the Armed Forces Medical Examiner called on him after American pilots died in the Gulf War.

The more I listened, the more Dr. Owsley sounded like the most famous person that I had never heard of. He was deeply involved—after the fact—in the most notorious mass disasters, wars, and crimes of contemporary times. But his role, and what he knew and saw, was shrouded in secrecy. When traditional forms of human identification—fingerprints, facial features, or clothing items—were absent, Owsley was brought in to identify people by looking at bones, from which he could ascertain a person's age, sex, race, and cause of death.

He honed his bone-reading skills at his day job as a curator for the Smithsonian's Museum of Natural History, where he was responsible for studying and preserving America's historic and prehistoric skeletons. In that capacity he uncovered crypts, visited tombs, excavated graves, and traveled to archaeology sites all over America and beyond.

Weeks after hearing about him, I visited Dr. Owsley at the Smithsonian. His office had all the earmarks of an ordinary science lab: X-ray lights, microscopes, magnifying glasses, and textbooks. But my eyes were drawn to the un-ordinary. The shelves along his walls were packed with what appeared to be oversize shoeboxes. Each one was nine inches wide, nine inches high, and thirty inches long.

"What are in these boxes?" I asked.

"Human remains," he replied.

The boxes seemed too small to hold a human skeleton.

"They will fit a human skeleton without crowding," he said, taking one off the shelf and removing the lid. It was my first glance at a real human skeleton. To me it looked like a collection of porous, brown-stained bones. To Owsley the skeleton was a human being, an intimate friend whom he knew almost as well as a family member. "You can learn more about a person from his bones than from anything else," he said.

Each box represented a treasure more valuable than gold or silver. The bones, he explained, were caches of knowledge.

He gave me a quick tour of the skeletons he was working on at the time: a recent homicide case, a Jamestown burial, a Plains Indian, a cowboy from Texas, some nineteenth-century pioneers, and a Civil War soldier. His work touched virtually every era of American history.

I felt as if I had slipped into a time warp and my escort was a real-life Indiana Jones. I asked him to elaborate on his job.

"I work in different kinds of worlds," Owsley began.

Those seven words convinced me that I was writing the wrong book. Rather than write about a lawsuit, I wanted to write about the man whose job is to visit other worlds, worlds of the past.

Over the ensuing three years I interviewed Owsley more than fifty times. I traveled with him to look at skeletons and mummies at various sites in the United States. I hiked with him in the West. I visited museums with him in the East. I observed him in his office and laboratory. The more I saw him in action as a professional, the more I wanted to know what made him tick.

At my invitation, he agreed to meet me in Lusk, Wyoming, the place of his childhood years. Armed with a video camera, a tape recorder, and a 35-mm camera, I had Owsley take me to the places of his childhood and introduce me to the adults who had most influenced him as a boy. I met his high school science teacher, his Sunday school teacher, and his Cub Scout den leader. They all pointed to his unquenchable curiosity as a boy. Each of them had felt early on that he would grow up to do something great, something most unusual, as a man.

Owsley and I even went to his childhood home and got permission from its current occupant to go inside and look around the basement where Owsley used to go to use his first chemistry set. I wanted to know everything about the climate and environment that helped produce Dr. Doug Owsley.

This book is not an anthropology textbook. Nor is it a scholarly treatise. It is instead my best attempt to show readers Doug Owsley's world through his own eyes. I hope that this book will give readers the feeling that they are in the Waco compound after the fire, in a Colonial crypt when the coffins are opened, or at the examination table when America's oldest mummy is unwrapped.

Arguably, no one in the world today has handled more skeletons than Owsley, more than 10,000 in all. He reads bones like most people read books. His vision is extremely rare. While we may see just a bone, he sees a story, a glimpse into the past.

In one of my numerous stays at Owsley's rural Virginia farmhouse, I sat at his dining-room table typing on my laptop. One evening, as I struggled to come up with words to convey to readers his ability to see things when most people would be in the dark, Owsley went outside to cut his grass. I got immersed in my writing, and hours later it struck me that Owsley had never come back inside. It was well after 10 P.M. It was pitch-black outside. Then I realized that the tractor motor was still running. He was cutting the grass in the dark.

The next morning I went outside, and the lines in the grass were straight.

Jeff Benedict
July 31, 2002

NO BONE UNTURNED

PROLOGUE

Summer 1962
Lusk, Wyoming

At five thousand feet above sea level and located just twenty miles west of the Nebraska line and not too far south of the Black Hills, Lusk was a child's paradise. Accessible by only two roads, the remote high-plains town with barely more than 1,500 residents was a throwback to the western cowboy era. Lusk had emerged during the gold rush to the Black Hills and, thanks to the building of the Cheyenne-Deadwood stagecoach route in 1876, had become a stopping point for freights of slat pork and whiskey, armored coaches bearing gold bricks, Indians, and opportunists from miners to armed bandits. In the early 1900s, silver was discovered in the nearby hills; but by the 1950s ranching had become the town's staple industry.

Although the stagecoach has long been supplanted by the Union Pacific Railroad, the ruts made more than one hundred years earlier still mark the ground, except for Main Street, now coated with asphalt, dotted with two stoplights, and lined with a row of storefronts. The commercial strip of local businesses includes a lumberyard, a bank, a diner, and a saloon. There are no chain stores and no fast food outlets. The Yellow Hotel, once a thriving two-story whorehouse for cowboys, and the Lusk Drive-in Theater, which abuts a corral, bookend the town. A

1

twelve-block-square grid making up the town's residential section sur-
rounds Main Street, an area of modest one-story ranch-style houses
with small green lawns lining streets with red fire hydrants and stop
signs on each corner.

The remnants of Lusk's past and its geographical isolation from the
rest of the world made Doug Owsley's childhood summers a never-
ending series of adventures. The curly blond-haired boy and Mike
Lyon, his dark-haired buddy with the crewcut, were as inseparable as
the explorers Lewis and Clark. Forbidden mines, fields littered with old
buffalo bones and arrowheads, and towering rock cliffs containing soil
and fossils dating back thousands of years were their playground. On
summer nights, you would sometimes find them sleeping outside,
under the carpet of stars in the vast Wyoming sky. During the day, they
would probe insect colonies, track roaming big game, and drink fresh
water from underground springs.

On a summer day in 1962, Doug and Mike visited one of their
favorite spots, the abandoned silver mine. Jumping off their bikes and
ditching them in the bushes, the ten-year-olds scrambled up the hillside
of the mine. Sand and pebbles trickled into their canvas sneakers as
they climbed higher. "Let's look here," said Doug, poking the ground
with a short stick as Mike paused, watching for the slightest movement.
The dusty soil camouflaged the brown-spotted horny toads they were
after.

"Let's go up further," Doug said, leading Mike toward the hill's
peak.

Both boys froze.

"Wow, look at that," Doug whispered. In a sloping circular pit, hundreds of sun-bleached white bones seemed to form the rough skeletal
shape of something too large to be human.

Doug jumped down into the hole amid the bones and rocks and
some broken pieces of a wood gate, its rusted metal hinges hanging
stiffly from the corners. "Man, this is neat," he said, picking up bones
and running his fingers over their dry, smooth surface.

"I don't know if I'd be touchin' those, Doug. They gotta be covered
with bugs."

Unfazed, Doug lifted the heavy, elongated skull. He put it right up to his face and peered into its empty cavities, recessed and separated by a long snout that jutted forward. The opening of the mouth had long teeth, with rows of shorter teeth on both sides of the jaws.

"That could be a horse," said Mike. Seeing that Doug had survived the initial contact with the bones, he finally climbed down into the hole and picked one up himself.

"I wonder how it got down here," Doug said.

"Must have fallen over the edge."

"What a discovery!" Doug said. "This is what it must be like to be famous archaeologists."

Mike wondered what the next step was. To Doug it was clear.

"This is a real treasure. We gotta bring this to our lab."

"What for?"

"Put it back together. . . . You know, like the dinosaurs in museums. Come on. Let's go get something to carry these in."

An hour later, they returned with their red metal wagons and filled them with armloads of the bones. After a series of trips to and from Mike's house, they had deposited the entire skeleton in their self-made science laboratory, a converted pigeon coop off of the Lyons' carport. Caged by gray chicken wire, the lab sported a scarred wood table and rickety shelves with glass jars holding frogs and insects. Some jars had masking tape on them, the words "For Scientific Research" scribbled on the strips.

A year earlier, they had made their first big discovery in the lab. Doug had dissected a frog and removed its thimble-size lungs for observation. Peering through the lens of Mike's microscope, Doug detected tiny worms working their way into the organs. It marked the first time they had seen live parasites.

The horse skeleton, however, took them to a new level of intrigue. Doug was sure he and Mike could reconstruct it. He spread the bones out on the table, sorting and matching those that looked like they fit together. He was prepared to spend the remainder of the summer, if need be, to assemble the skeleton.

• • •

The wooden sign staked in the ground proclaimed GAME WARDEN STA-
TION and held a blue-and-yellow insignia of an antelope accompanied
by the words "Wyoming Game and Fish." Behind the sign sat a cozy
ranch-style redbrick house with a narrow concrete driveway. Inside,
thirty-five-year-old Norma Lou Owsley wore an apron as she prepared
dinner at the kitchen counter. In the adjacent living room, the brass
pendulum on the wall clock swung, its gentle ticking a sound that
Norma Lou had grown accustomed to since moving into the house on
South Lynn Street in 1958. That year, her husband, Bill, had become a
Wyoming game warden. For monitoring all hunting and fishing in the
county, as well as counting herds and feeding elk in the Yellowstone
area in the winter, Bill Owsley was paid $195 a month. The state also
provided him with a home for $10 a month. The Owsleys lived conser-
vatively, and they did not smoke or drink, yet they never made it past
the third week of the month before running out of money. It was
Norma Lou's job to make the food stretch. She did it by making lots of
casseroles, including Doug's favorite: diced onions and celery with
ground beef and cream of mushroom soup over rice.

Bill practically lived in his pickup truck, his game warden duties
keeping him away from home from dawn until dark. That left Norma
Lou to help Doug with his homework every night, usher him to Cub
Scouts once a week, and take him to church on Sundays. When it came
to schoolwork, she never had to prod Doug. Bill promised him a quarter for every A. The challenge stoked Doug's competitive nature, moti-
vating him to study, which consistently put him at the top of his class.
He treated scouting as a competition too. He always earned the most
badges, found the most bugs, and identified the most types of leaves.
He grew insatiably curious. During recess his Sunday school teacher
would often find him crouched over, his elbows on his knees, searching
for insects in the churchyard grass and observing their actions while the
other children played games.

Unlike some game warden residences, the walls of the Owsley home
did not display the stuffed heads of wild bear, coyote, or deer mounted on
wooden plaques. Bill rarely hunted and disapproved of making trophies of
animals. And Norma Lou, an accomplished artist, preferred her paintings

and cut glass. The Owsleys instilled in Doug a respect for animals. Yet he looked at animals in a different light—they were specimens whose anatomy and biology held a deep fascination for him.

When he was nine years old, he began experimenting in the basement with a chemistry set, a Christmas gift from his parents. Combining chemicals from the kit with bathroom solvents, he developed an anesthetic to knock out frogs by applying the compound to their nose with a cloth. One day, with a frog unconscious, Doug slit open its belly to observe its heart and other organs. The burgundy red heart no bigger than a pencil eraser pulsed while the pink, porous lungs expanded and contracted. Fascinated, he took notes, then stitched the frog up with one of his mother's sewing needles. When it revived, Doug palmed it in his hand and headed up the basement steps and through the living room en route to its home, a nearby pond, where he planned to release it. Emerging from the kitchen, Doug's mother intercepted him.

"What are you doing?" she asked, suspecting he was up to something.

Knowing his mom would disapprove, Doug dropped his hand to his side. But the frog poked its head out between his thumb and index finger.

"What do you have?" she asked.

"It's a frog I'm taking outside." The less said, the better, Doug thought.

Overhearing the conversation, Bill rose from his chair in the living room and approached. "Let me see it."

Dreading a stern lecture, Doug opened his hand slowly. The frog was on its back, and Bill immediately noticed the stitches running up the frog's belly.

"No more of that," he said, a man of few words. "That's cruel."

Deflated, Doug left the house. His parents just didn't understand. He wasn't trying to be cruel; he just wanted to know how the frog's vital organs worked. It was all for science!

From that day forward, as far as his parents were concerned, at least, he stopped dissecting living frogs in his basement. That was also when he moved his laboratory to Mike Lyon's pigeon coop.

• • •

It was the last week of summer. For two months Doug, with Mike's help, had been rebuilding the horse skeleton they had discovered. Without the aid of a textbook or diagrams, Doug had paired bones of similar sizes and shapes, then fit them together to form a partially articulated skeleton. To him, the bones were a giant jigsaw puzzle. The more he worked on it, the more the animal became whole. Mike stood by him every step of the way. It was the coolest thing, being real-life archaeologists.

"When we're scientists someday," Doug said later, when the horse skeleton was complete except for a few odd or missing bones, "I'll bet we could be in *National Geographic.*"

Eleven years later,
two hundred miles away

A twenty-one-year-old senior at the University of Wyoming, Doug climbed into the passenger side of a brown 1971 Ford pickup truck. It belonged to his anthropology professor, George Gill. Ten years older than Doug and a Vietnam veteran, Gill looked like a cross between the Marlboro Man and Robert Redford. He wore a cowboy hat over his dark bushy hair, and a sleeveless khaki field shirt that displayed a suntanned, muscular chest and chiseled biceps. Gill's empty gun rack rattled against the rear window as he and Doug drove off, headed toward Pitchfork Cave in the Absaroka Range. Weeks earlier one of Gill's students had been rock climbing in that region and spotted what looked like a human skull. Gill reported the discovery to the Wyoming State Archaeologist's office, and they in turn gave him authorization to recover it.

Gill had become a mentor to Doug after meeting him just one year earlier in his Introduction to Anthropology course. On the first day of class Gill could not help but notice the young man. Many of his students had long hair, wore tie-dyed shirts, avoided the front row, and never asked questions. Doug wore dress slacks and a collared shirt, and his hair was neatly trimmed above his ears. He sat in the center seat of the first row, where he regularly asked questions that sometimes elicited snickers from his classmates. Snicker as they might, Doug was the only student who earned a perfect score on the first exam. In fact, he would go on to ace every exam Gill offered that semester, finishing far ahead of his peers. Curious, Gill checked the student records and discovered that through Doug's first three years of college, he had a 4.0 grade point average. According to the files, he was premed, on track to go to medical school and become a physician.

As the semester wore on, Doug started spending his spare time in Gill's lab, one wall of which contained shelves lined with bones, skulls, and casts of skulls. Each of the four shelves was dedicated to one of the stages of human evolution: Australopithecine (earliest definite hominids, believed in existence in regions of Africa between 4 million and 1 million years ago); *Homo erectus* (hominids from Java, China, and Kenya, dating back between 1 million and 500,000 years ago); Neanderthal (hominids found in Europe and western Asia, dating back between 200,000 and 40,000 years ago); and modern *Homo sapiens* (early modern human, which first surfaced 100,000 years ago).

By semester's end, Gill could remove any bone or skull from a shelf and Doug could tell him which period it came from, based on its shape and features. Gill encouraged him to take a couple of graduate-level courses during his senior year, human evolution and osteology, a skeletal-biology class. Doug would go on to finish first in both classes, outperforming the graduate students. Gill was taken with him. The kid was a natural. But Gill didn't want to push him into anthropology. Doug needed to realize it on his own.

Before the end of his osteology class, Doug submitted to Gill a handwritten thirty-page paper entitled "The Criteria for Siding Hand Bones." He told Gill he had done it for extra credit.

Gill flipped through the pages. Doug had spent weeks handling and comparing tiny carpal bones and had produced a text on how to distinguish left-hand bones from right-hand bones.

"It's just amazing," Gill said.

"What is?"

Gill smiled. "Doug, you certainly don't need ten extra points. You've got the highest grade average in the class."

"This stuff is more exciting than medicine," Doug said. "I think maybe this is the field I should be in."

Gill was thrilled. Doug had rare abilities and gifts. As an undergraduate he was making observations and deductions that matched those of Gill and his colleagues. His thirst for knowledge was unquenchable, and he had a phenomenal rate of retention. "You would do extremely well in this field," Gill said.

Doug seemed doubtful though. After all, he was a premed major.

"Let me tell you something, Doug. I was a zoology major and heading down the premed route myself. Heck, half of us physical anthropologists were premed majors." Gill told him that the best premed students often desert medicine for physical anthropology, a field that offers the freedom to travel and learn.

Doug's enthusiasm prompted Gill to bring Doug to Pitchfork Cave, providing an opportunity for him to discover and handle his first human skeleton.

Gill slowed his truck to a stop beneath the cave high up in the face of a cliff. After stepping out of the vehicle and surveying the precipice, Gill and Doug pulled up the extension ladder and secured it against the rock wall. Doug trailed Gill up the ladder. Cresting the ledge, he came to a dead stop.

Just feet away, a face peered back at him from out of the rocks. Speechless, Doug stared at the skull. Immediately, his mind began racing, wondering who that person once was.

"Wow, this is fantastic," Gill said after he pulled himself up onto the rock ledge. Unlike most people, anthropologists are excited by the appearance of a skeleton, welcome it in fact. For them it is a mystery waiting to be solved, a window into the distant past.

Gill told Doug to get the tools from the truck. Doug hustled down the ladder and came back moments later with hand trowels, dental picks, small paintbrushes, whisk brooms, and storage bags. Gill was already taking photographs. Doug gently put down the tools, as if in a sacred place, and approached the small stack of limestone rocks sur-rounding the skull. Squatting beside it, he saw its lower jawbone and teeth, visible as before through a gap in the rock pile. The left side of its face was completely skeletonized.

Carefully, Gill pulled away the rocks, and the rest of the skull grad-ually came into view. Doug was astonished at how well preserved it was. The right side of the face still held skin tissue. Two round, flat cop-per earrings hung from the still-intact right ear, while braided black hair ran down each side of the head. The hair on the back of the head was loose but matted.

Following Gill's lead, Doug used a paintbrush to gently brush back dirt, exposing the torso, ribs, and a leg. Suddenly they spotted a second skeleton—this one without a skull. Gill stressed to Doug the need to fully expose the skeletons, and the importance of then photographing and documenting their position.

Mirroring Gill, Doug used a trowel to remove sections of the grave where the dirt was compacted, then returned to using a brush to remove the windblown soil that covered the majority of the skeleton. It took less than two hours to uncover the burial.

Both skeletons were wrapped in buffalo robes, lying face up in a shallow grave that stretched ten feet in length and no more than two inches in depth. The grave had previously been disturbed, leaving por-tions of the skeletons exposed and missing. Most of the limbs of the first skeleton and the skull of the second one were gone.

The missing skull nagged Doug. "Where is that other skull?" he asked.

Given the height of the cave's opening, Gill figured an eagle or some other large bird had probably dislodged it from the grave.

The more Doug thought about it, the more it bothered him. He dug around the cave looking for it, then climbed back down the ladder and hunted around beneath the cave. No luck. He wanted to keep searching, but Gill suggested they deal with the bones at hand, pointing out to Doug the need to identify both skeletons.

Gill took a five-by-eleven-inch card, marked it "PITCHFORK Burial #1A," and propped it against the skeleton with a skull. He put one marked "PITCHFORK Burial #1B" next to the skeleton without a skull. After taking slide pictures of the grave and the positioning of the skeletons, he examined the specimens more closely.

"What do you think the sex is?" Gill asked.

Doug studied the heavy brow ridges and the cranium. It seemed obvious to him. "Male," Doug said, glancing up at Gill.

"That's right. And the race?"

Both skeletons were big and robust. The nose shape on the one skull was distinctive, like an upside-down heart. It had a very long mid face and very prominent, large cheekbones. "Indian."

Gill nodded, impressed but not surprised. They next turned their attention to the necklaces made of hundreds of glass beads—red, purple, white, blue, and black. Some had come loose, but most of the beads were still on a thin brown sinew, made from animal tissue. Gill handed Doug a couple of small paper bags and small plastic vials with snap-on tops. "It's important to keep these intact when you place them in the bags," he said. The patterns and stitching held clues to tribal affiliation.

Doug used his fingers and a dental pick to pick up beads, meticulous in his attempt to preserve them in the precise order in which they appeared in the grave. When he finished, he crouched beside Gill, who was peeling back the clothing over one skeleton's chest. It had on several layers of robe that appeared made of deer or buffalo hides. The edges were decayed. Gill told Doug to note how many layers they went through.

Doug nodded.

Beneath the robes, the skeleton had on a faded red military coat with a blue lining, metallic trim, and brass buttons. "Looks like an old U.S. Cavalry jacket or something," Gill said.

The chest was still held together by ligaments, Doug observed. The hand bones had some cartilage as well.

Gill stopped to explain. Anytime a skeleton retains some type of soft tissue it can be classified as a mummy, or at least partially mummified. These skeletons had been well preserved, as a body can completely skeletonize, without any sign of skin, internal organs, or even brain tissue. Brain tissue is usually one of the last things to go because it can tan itself as it shrinks. It is often reduced to a little leathery piece of material rattling around inside a skull.

Feverishly scribbling notes, Doug looked up as Gill pointed to an area of the skeletons where soft tissue was visible. A skeleton inside of clothing, he explained, will preserve better. Gill pointed to the ear with the earrings, the hair behind the head, and the skin around the skull and the torso. The line that divides a skeleton from a mummy is not a real clear one, Gill explained. They had found a partially mummified skeleton.

Doug also noticed tiny, empty insect casings around the mummy portion of the skeleton. They were casings that fly pupae shed, Gill explained. They were a clue to what time of year the two men were buried. These bodies had to have been interred at a time when the flies were out, ruling out certain months of the year.

After completing the recovery, Doug helped Gill put the remains in boxes and load them into the truck for safe transport to the state archaeologist's office. Doug wondered out loud if any Indian wars had taken place in the region. Gill said that folklore and some historical documents suggested that there had been a number of battles around Pitchfork Ranch between the Blackfeet, Crow, and Shoshone Indians.

The mystery surrounding both men captivated Doug's thoughts for weeks afterward. What were two Indians doing up in a cave, dressed in military uniforms? Did both men die simultaneously? And how did they die? Neither skeleton showed signs of trauma. It irked him that he didn't have time to pursue the questions.

Stirred by what he saw, he yearned to know more about American Indians, their lifestyle and history. Inevitably, he lost interest in medical school and accepted a teaching assistantship to study under Dr. William Bass, recognized as the country's top forensic anthropologist. It was too good an opportunity for Doug to turn down. Prior to becoming chair of the University of Tennessee's anthropology department, Bass had personally recovered more American Indian remains than any other anthropologist in the United States.

After World War II, the federal government had initiated a vast public works program that included nationwide dam construction along the country's major rivers. Prior and during construction the Interior Department conducted archaeological surveys in the river regions. The Smithsonian Institution and the National Park Service worked together to salvage archaeological sites detected in the surveys. The Smithsonian asked Bass to assist in the Missouri Basin Project—a massive retrieval of artifacts and human remains sure to be exposed with the construction of four reservoirs along the Missouri River in North and South Dakota. With the river dammed in four locations, water buildup caused wave action and subsequent erosion, washing out old Indian villages and burial grounds.

Hoping to salvage the sites on behalf of the Smithsonian Institution, Bass spent fourteen summers from 1956 to 1970 excavating and recovering human remains along the Missouri River. With the aid of local tribes, he found approximately five thousand Plains Indian skeletons. One of the locations he dug was the Larson site, a fortified village in South Dakota that was home to the Arikara Indians prior to 1780. The site contained earth lodges—domed huts with sod roofs that were supported by timbers—and a tribal burial ground. He excavated approximately seven hundred burials at the cemetery and also acquired sixty-five individuals' remains in and around the village homes.

Shortly after Doug started graduate school, Bass directed him to construct a demographic profile of the Arikara. While studying forensic anthropology under Bass, Owsley determined the age and sex of 762 Arikara skeletons. He found that 40 percent of the individuals were newborn babies who had died at birth or shortly after. He also discovered many skeletons of females between the ages of fifteen and nineteen, indicating an earlier average age of death for females than males. The data became the basis for Owsley's master's thesis: a comparison between the mortality rates of males and females in the Arikara Indians at Larson Village, and an attempt to explain the discrepancy.

After completing his research, Doug met with Bass to discuss his findings. He had discovered that the skeletons recovered from the village had never been formally buried; rather, they were found scattered on the house floors and in surrounding areas. The lodges had simply caved in on top of the remains. Many of those skeletons were males and older women. The village contained relatively few children and young adult women.

On the other hand, the skeletons recovered from the cemetery had generally been formally interred in the flex or fetal position. These skeletons contained a higher proportion of young adult women and infants.

Statistically, the age and sex distributions of the skeletons at the cemetery differed from those in the village. Owsley wondered why. Bass figured a catastrophic event—most likely a smallpox epidemic—had struck the village. Historical accounts from fur traders in the Missouri Basin referred to the practice of abandoning smallpox victims in their homes and deserting the village in order to prevent the spread of disease. Some villages were even burned. The Larson Village seemed to fit the profile. One Arikara home in the village contained forty-four individuals. It and other homes had also been burned, as some skeletons showed signs of charring.

Doug had also found that skeletons buried in the cemetery were of individuals who had not all died instantly, but had been buried over the course of a few decades. This could be explained by Bass's belief that smallpox had killed those in the village, preventing them from receiving a formal burial.

With the thesis complete, Bass asked Doug to help him teach his Introduction to Physical Anthropology class. Early in the semester, Bass brought in a box of nine skulls, planning to use them as teaching aids when going over the anatomy of the skull. Some of the skulls were from the Larson site that Doug had studied for his thesis.

As Bass lectured, Doug sat at the table on which the box of skulls rested. His mind wandering, he found himself staring at them, then noticed a fine incision on the front of one of the skulls. He picked it up and examined it more closely. Another incision ran along the side of the skull. Cut marks, he thought.

Recognizing the skull as one from the Larson Village collection, Doug looked in the box for others. There were three more. Doug quietly lifted the other three Larson skulls from the box. All three had similar cut marks around the tops of the cranium. They looked as if they had been scalped.

Doug had just completed a term paper for his archaeology class on evidence of scalping in the southeastern United States. In researching how Indian tribes scalped enemies—removing the hair and soft tissue on the top and back of the head with a blade—Doug had read scholarly studies that detailed the process. One scholar reported: "They with one hand twisted in the hair, extend it as far as they can [and] with the other hand . . . speedily draw their long sharp-pointed scalping knife, give a slash round the top of the skull and with a few dexterous scoops soon strip it off."

Prior to European contact, tribes had used sharp reeds, shells, and flint knives for scalping. The method became more efficient after Europeans introduced steel-blade knives to the plains, enabling Indians to complete a scalping with only two cuts to the occipital bone. To learn what scalping cuts looked like, Doug had traveled to Vanderbilt University and examined scalping victims' skulls recovered from prehistoric Indian burials in Tennessee. The skulls contained cuts like the ones Doug saw on the Larson Village skulls. He wondered how he and Bass had not previously detected the cut marks.

As soon as Bass finished his lecture, Doug hustled up to him.

"Dr. Bass, I have to show you this," he said, pulling skulls from the box.

He pointed to the cuts on the skulls.

Bass looked closely at the incisions. He and several of his doctoral students had been handling the Larson skeletons for more than a decade and had never noticed the cuts. They hadn't been looking for incisions, but instead had been taking measurements.

"I've seen examples of this," Doug continued, reporting on his term paper.

"Well, Doug, you need to document all of this," Bass said, encouraging him to reexamine the entire Larson collection for evidence of cut marks on other skulls.

This time, instead of looking for indicators of age and sex, Doug looked for evidence of trauma and mutilation. The more he looked, the more he doubted the conclusion he had reached in his thesis paper, that an infectious disease such as smallpox had decimated the Arikara. Virtually every skull recovered from the village area contained scalping cuts. The only skeletons from the village that didn't show signs of scalping were ones that had been burned or decapitated. On the ones that had been decapitated, Doug found cut marks on the vertebrae of the neck.

The trauma explained why skeletons were found randomly scattered around the village area and inside the houses, rather than buried with the ones in the cemetery. The Arikara village had been violently attacked. The catastrophe that instantly wiped them out was physical violence, not disease. In addition, their homes had been burned by the invaders, a conclusion that helped explain why Doug found so few young adult women in the village as compared to the cemetery. After killing the Arikara men, the enemy likely took captive the young adult women, while leaving behind the older women and young children. Doug attributed the violence to an enemy tribe. Dated between the late 1600s and 1750, Larson Village predated European contact in the South Dakota region.

Bass was so impressed with Doug's discovery that he sent him to present his research to a group of professional archaeologists and university professors at an archaeology conference in Lincoln, Nebraska. Doug had never presented at a professional conference and had virtually no experience with public speaking. Facing two hundred people, he used an array of slides while telling a chilling tale of how another tribe butchered the Arikara.

Most of the skulls, he explained, were scalped, even though some of them were also burned. "There were three or four males that had neck vertebrae that were decapitated." On the slide projector, he put up a slide showing an example. "You'll note the obvious cut marks on the higher cervical vertebrae," he said, applying his pointer to the screen.

He put up another slide, then turned to his audience. "Basically, they cut from ear to ear," he said, demonstrating by moving his thumb in a quick jerking motion from his right ear, across his neck to his left ear. "And deeply into the bone." With his right hand, Doug reached up and grabbed a handful of his hair and yanked his scalp back. "Then they pulled the head back so forcefully that it snapped the second cervical vertebrae, causing the head to pop off."

As he spoke Doug noticed that the audience remained silent. He wondered if they were captivated, or bored.

He put up another slide. "This one is of a young woman," he said, his speech speeding up. "She has been dismembered, obviously. They cut across her radius and ulna." Doug held up his left arm. Using his right hand, he sliced across his left wrist, as if he were cutting with a knife. "And then they grabbed the hand and pushed it back so forcefully," he continued, pulling his left wrist backward, "that it was taken off."

People in the audience winced. "In the literature it is clear that not only heads or scalps were taken as trophies, but so were hands."

Doug shut off the slide projector and turned on the lights. Silence stifled the room. The largely academic audience was unaccustomed to such graphic evidence of violence. "We see what we are trained to see," Doug concluded softly, admitting to his audience that he had originally assumed that the Arikara village had been wiped out by smallpox. He had approached his master's thesis research to fit that conclusion. "We have to be able to step back and open our eyes more broadly and focus on greater details. We have to make the conclusion fit the data, not the other way around."

The professors broke into applause, as if listening to a distinguished colleague. Yet Doug was the youngest presenter at the conference and had not completed his Ph.D. When he returned to Knoxville and reported on the conference, Bass grinned.

Doug had one question. "Do you think I'll be able to find a job when I'm through with my degree?"

Bass smiled. He had never taught anyone as driven or talented as Doug. He had a feeling that Doug Owsley was going places.

1

HIGH-STAKES PLAYGROUND

March 26, 1992
Washington, D.C.

Against a backdrop of Gothic columns and massive granite walls, an elephant-size ivory statue of a brontosaurus stood erect in front of the Smithsonian Institution's National Museum of Natural History. Hustling down Constitution Avenue, forty-year-old Doug Owsley glanced at the dinosaur before entering the museum's gold-encased doors. He flashed his ID and nodded as he passed a security guard and headed to a staircase in the far corner of the lobby. Wearing a jammed backpack and a windbreaker, he took a staircase beyond the exhibits, climbing all the way to the top floor of the museum, an area off-limits to the public. He unlocked a secured door and entered a labyrinth of tall metal cabinets. They contained the Terry Collection, more than 1,600 cadavers collected by anatomist Robert Terry for training purposes to the St. Louis University Medical School, which were later donated to the Smithsonian.

Owsley passed the cabinets and turned down a cavernous, dimly lit hallway. Both sides were lined from floor to ceiling with cast-iron shelves holding green wooden boxes filled with skeletons from infants to adults. In all, the Smithsonian housed more than thirty thousand skeletons, more than any other institution in the United States. Its collections included remains from the Colonial period, nineteenth-century farmers, American Indians, African Americans, soldiers from the Civil War and Custer's army, Chinese immigrants who had worked in Alaska, and the famous Huntington Series, an anatomical collection of three thousand European immigrants who had died in New York City.

After Owsley graduated from Tennessee with his Ph.D. in 1978, he taught part-time there for one year while job hunting. During that time, he continued to hone his forensic skills by going to crime scenes with Bass to identify and recover bodies. He also did postdoctoral studies on the skeletal collections of American Indian remains Bass had recovered from the Midwest. In 1980 he took a full-time teaching position at Louisiana State University. During the summers he returned to Tennessee to continue his research on American Indians. Doug had been at LSU for five years when he heard that the Smithsonian's Museum of Natural History was looking for a curator to take over responsibility for its vast collection of American Indian remains. Convinced that the Smithsonian would attract a who's who of applicants from the field of anthropology, Doug was not going to bother applying. He had had his Ph.D. for only a few years and had held only one job. His credentials, he felt, would be inadequate to satisfy the Smithsonian, an institution that had obtained almost mythic status in his mind.

Congress passed legislation creating the Smithsonian Institution in 1846 after English scientist James Smithson left his fortune—over four hundred thousand dollars—to the people of the United States. In his will, Smithson, an internationally recognized chemist who died in 1829, specified that the money be used to found an institution in Washington "for the increase and diffusion of knowledge."

His gift triggered a controversy. Some members of Congress argued that it was inappropriate for the United States to accept a gift from a British nobleman and name an institution after him. President Andrew Jackson disagreed and felt Americans would benefit from such an institution. He asked Congress to pass legislation authorizing the acceptance of the gift. Massachusetts congressman John Quincy Adams, the former U.S. president who was chairman of the congressional select committee charged with reviewing Smithson's gift, agreed. In 1836

Adams drafted legislation authorizing Congress to take possession of Smithson's fortune and to create a national museum. He wrote:

> If, then, the Smithsonian Institution, under the smile of an approving Providence, and by the faithful and permanent application of the means furnished by its founder to the purpose for which he has bestowed them, should prove effective to their promotion; if they should contribute essentially to the increase and diffusion of knowledge among men, to what higher or nobler object could this generous and splendid donation have been devoted?

Ten years later, in 1846, the Smithsonian was founded. Today it is the world's largest museum complex, with more than sixteen museums, four research centers, the National Zoo, and vast libraries. When Doug told Bass about the curator position at the Institution's Museum of Natural History, Bass insisted that he apply.

"The Smithsonian would be the perfect place for you," Bass told him. "It has great collections and it's a great place to be in terms of forensics. You'll be called to help the FBI and other law enforcement agencies identify bodies. You'll see it all. The Smithsonian is a high-stakes playground."

Bass's connection to the Smithsonian dated back to 1956, with the Missouri River and Dakotas excavations, a project that had stretched over fourteen years. A number of Bass's students had gone on to become some of the Smithsonian's top curators. But none of Bass's students had impressed him as much as Owsley. "Trust me," he told Doug, "you've got the knowledge to work there."

Bass recommended Doug to the Museum of Natural History, noting that besides his academic credentials, Doug's experience handling skeletons was unparalleled. A skeletal biologist with advanced training in human anatomy and forensics, Owsley had worked on over two thousand human bodies from archaeological sites, morgues, crime scenes, graveyards, and battlefields.

In 1987 Doug was shocked to learn that he had been chosen to be a curator in the museum's Department of Anthropology. His chief responsibility was the preservation and study of the museum's vast collection of North American Indian remains. His work at Tennessee and LSU had made him an expert on the skeletal biology of Plains Indians, focusing primarily on their health conditions and the mortality rates of women and children. In his first few years at the Smithsonian he would publish a landmark textbook about the migration, warfare, and health and subsistence practices of Plains Indians.

His experience working with skeletal remains at modern crime scenes had also caused his star to rise fast at the Smithsonian. As a federal agency partly funded by Congress, the Smithsonian encouraged its scientists to aid other federal agencies, a tradition that dated back to 1938. That year, FBI director J. Edgar Hoover wrote to the secretary of the Smithsonian requesting assistance from Dr. Aleš Hrdličvka, the museum's first curator of physical anthropology and the founder of the American Association of Physical Anthropologists. Hoover wanted Hrdličvka's assistance in evaluating specimens thought to be human. Back then the country had few scientists qualified to study skeletons for clues to crimes, and the FBI's offices were across the street from the Smithsonian. Hrdličvka went on to report on thirty-seven cases for the FBI, determining whether remains were human or nonhuman, and estimating the sex, age at death, ancestry, living stature, and time since death.

After Hrdličvka died in 1943, the use of Smithsonian anthropologists as consultants to the FBI increased. By the time Owsley joined the Smithsonian, his predecessors had assisted the Bureau on over eight hundred cases. In addition to doing some work for the FBI, Owsley got requests from other federal agencies, such as the State Department, the ATF, and the Park Service, as well as branches of the military. When unidentified bodies surfaced or when the cause of death needed to be ascertained, the government increasingly turned to Doug.

Halfway down the hall he stopped at an oversize brown metal door with a large double-paned window. A paper sign taped to the door read, "Room #345 Dr. D. W. Owsley." Inside, the observation tables and countertops were cluttered with human teeth, bones, microscopes, X-ray sheets, and dental picks. In the corner stood a china cabinet full of skulls. He went directly to his private office adjacent to his lab.

His computer screen off and a stack of anthropology and anatomy books and neglected correspondence littering his desk, Owsley stared at a State Department telegram marked CONFIDENTIAL. The U.S. embassy in Guatemala had sent it to the secretary of state's office back in 1985, confirming the disappearance of two American journalists who had ventured into the Guatemala highlands to interview left-wing guerrillas entrenched in a three-decades-long civil war against the Guatemalan government.

Rubbing his eyes and adjusting his wire-rim glasses, Owsley picked up a copy of the telegram and began reading.

> Two EMBOFFS [embassy officials] traveled by helicopter on April 18, 1985, to the [provinces] of Huehuetenango and El Quiche [Quiché] in an investigation of the welfare and whereabouts of two missing AMCITS [American citizens], Nicholas Blake and Griffith William Davis. The missing AMCITS were not found, but their movements up to March 29 or 30 were traced and verified. Given the dangerous nature of the area where Davis and Blake may have gone, Embassy does not plan to send staff into the field further to look for them.
>
> The area is rugged high sierra, with few roads and an active guerrilla insurgency. The route that Blake and Davis planned to hike is along the northern face of Los Cuchumatanes, the highest mountain range in Central America. The elevations along the route vary from about 6,000 to 9,000 feet above sea level. The entire area is included in the Embassy's Travel Advisory, which identifies places which are not considered safe for tourist travel because of frequent clashes between the guerrillas and GOG [Guatemalan army] Security Forces.
>
> It would appear that Davis and Blake . . . did not go to Salquil where they were told they could go. It is possible, instead, that Davis and Blake took the path to the guerrilla-controlled village of Sumal, where they were told not to go. If this is what Davis and Blake did, it must be assumed that they are now out of contact because they do not wish to be found, are being held captive by the EGP, or they are dead. End Summary.

Owsley put the telegram on his desk. The request to find and identify the missing remains of photographer Griffith Davis and freelance writer Nicholas Blake came directly from Blake's brothers Randy and Sam, who had been searching for their brother for seven years. When the initial investigations by the U.S. embassy failed to turn up their brother and Griffith Davis or the cause of their disappearance, the Blake brothers aggressively lobbied the embassy and other agencies in the United States government to keep looking. Their missing brother Nick knew Vice President George Bush's daughter Dorothy, and they appealed through her to the vice president for help. On their behalf, Bush directly called Guatemalan military dictator Mejia Victores and requested his personal intervention in the case.

Over the next two years, no solid leads materialized until the Blakes established contact with a Guatemalan schoolteacher from the village where Blake and Davis were last seen alive. The schoolteacher gave the Blakes a chilling report. He said that in March 1985 Blake and Davis had spent the night in a schoolhouse in the village of El Llano, an area policed by a ruthless paramilitary force—civil patrols—charged with doing the army's dirty work, from identifying guerrillas to killing suspected sympathizers. The following morning, five or six civil patrol members led Blake and Davis out of the village and shot them. Details on motive were slim. But the Blake brothers suspected that the army, which issued orders to the paramilitary civil patrols, had a role. Nick had previously gone to the highlands and exposed other atrocities committed by Guatemala's military regime. And the army knew he was back in the war-ravaged area with a photographer.

The bodies were dumped in some brush and covered with logs. As much as two years later, villagers were ordered by the Guatemalan military to move the bones one kilometer from the murder scene, where they were to be broken up and burned. The schoolteacher reported that the bones were destroyed out of fear that the United States would retaliate not only by killing the persons responsible for the murders, but by wiping out the entire population of El Llano. Burning, they hoped, would remove all traces of the crime.

By the time the schoolteacher's report reached the Blakes, George Bush had been elected president, and the U.S. embassy reactivated its investigation. And early in 1992, the Blake brothers reached a private agreement with Felipe Alva, a regional commander who oversaw forty thousand of Guatemala's approximately nine-hundred-thousand-member paramilitary forces. Convinced that men under Alva's command were responsible for the killings, the Blakes nonetheless agreed to pay him ten thousand dollars when Alva claimed he could deliver the human remains. The Blakes made a partial payment up front, with the balance contingent on the bones being positively identified as those of their brother Nick and his friend Griffith Davis. They also assured Alva that they would not press international criminal charges against Alva and his men. They just wanted their brother's remains.

In March 1992, Alva notified the Blakes that he had recovered two crates of remains. Hearing that Owsley was their best bet to identify the bones, the Blakes called him. They had to give him some background on the situation in Guatemala. Seven years had passed since their brother and his colleague had disappeared. The Guatemalan government blamed their disappearance on the Cuban-backed Communist guerrillas whom Blake and Davis had gone to interview. Since then, the rebel forces had been defeated by Guatemala's military regime. In the process of rooting out the guerrillas, Guatemala's military killed or caused to disappear an estimated two hundred thousand civilians and destroyed roughly 650 villages. Hundreds of thousands of civilians were displaced from their homes and forced to live as refugees. Guatemala's military tactics during the late 1970s and 1980s were so brutal that the U.S. Congress eventually cut off foreign aid to the government and the State Department placed travel restrictions on Americans desiring to visit Guatemala.

Owsley agreed to work with the Blakes and helped them obtain from the U.S. Department of Agriculture a permit to bring foreign soil into the United States. Sam Blake went to Guatemala and transported the remains to Washington. Which brought him to today. The bones had arrived.

Owsley's phone rang.

"Dr. Owsley, you have some visitors here to see you," a security guard said. "Randy and Sam Blake. They've got some boxes for you."

"I'll be right down."

Moments later, Owsley, in his white lab coat and khaki pants, approached the loading dock behind the museum. Reaching the security guard post, he spotted two rectangular wooden boxes that resembled coffins. Randy and Sam stood beside them, visitor badges in hand.

"Hi, Dr. Owsley," Randy said, offering his hand and introducing his younger brother Sam, a special assistant to the director of the Pentagon-funded National Security Program at Harvard's Kennedy School of Government.

"Nice to meet you," Owsley said.

"This is it," said Randy, patting the top of the boxes.

"Well, let's go inside and take a look," said Owsley.

In Owsley's laboratory, the Blakes placed the crates on a metal observation table. Owsley pulled out a release form and on the line next to the heading "Human skeletal remains and/or material evidence consisting of" he wrote, "Two boxes with soil and possible human remains recovered from a site in Guatemala."

He assigned Smithsonian Case No. 92-3 to the form, and Sam signed it.

Setting the form aside, Owsley gently pried open the two boxes, both of which were lined in violet velvet. Each box contained seventy-five pounds of dirt with lumps of red clay and tissue-thin pieces of gray ash. With his bare hand, Owsley reached into the boxes, plucking out the largest items he could feel: four metal tent stakes, a bone socket approximately two inches in length, some tiny charred bone fragments, and a tooth. One by one, he placed them on the box top.

Other than the one bone socket, the boxes contained no bone fragments bigger than a fifty-cent piece. "These are fully cremated bones, long past the charred stage," he said. He noted that the bone fragments were in a calcined stage, wherein the organic components of the bones—cells and proteins—had been destroyed, leaving only the mineral portion of the bone. "This is just bone mineral," Owsley said, holding one fragment up to show the Blakes. "These bones were exposed to a high-intensity fire."

Randy picked up a tiny bone fragment and looked at it, pursing his lips, while Sam kept his hands in his pockets and said nothing. The scarcity of bones, combined with their charred and cremated status, left scant evidence for Owsley to work with. Randy and Sam seemed deflated.

The challenge only provoked Owsley, convinced he could solve the identity mystery with a few more clues.

"Don't worry," Owsley said, trying to rally them. "We'll figure it out. People put their victims through wood chippers and burn them and we still figure it out."

Before the Blakes left his office, Owsley provided them with a list of additional information he would need to help identify the remains:

whether Nick and Griffith were packing a tent

what kind of clothing they might have been wearing

what, if any, medical X rays existed on both individuals.

The Blakes gave Owsley the phone number for Griffith's mother, Dolores Davis, and suggested he call her about X rays for Griffith. The Blakes had maintained contact with Ms. Davis since the disappearance, keeping her apprised of new leads and developments. She remained hopeful that her son would one day be found—alive.

For the next two weeks, Owsley thoroughly examined the entire contents of the two boxes. Using a one-eighth-inch wire mesh screen, he sifted out all the larger items from the soil: plant roots, charred wood, and 1,610 tiny bone fragments, none of which exceeded an inch in length. Two bones, each about the size of a dime, stood out. Owsley fit them together, partially forming the right frontal sinus, a small bone cavity in the center of the forehead, about an inch above the nose. When he X-rayed the two pieces together, he noticed an irregular groove running through the two bone fragments. The groove was well defined and appeared to run from the top of the nose toward the hairline. Unable to completely reconstruct the right frontal sinus without the missing third bone fragment, Owsley crossed his fingers, hoping that head X rays existed for either Blake or Davis that would confirm whether one of the men possessed an irregular groove.

While Owsley analyzed the bones in his lab, the *New York Times* ran a headline story, "Progress Is Seen in Guatemala Case," reporting that the recovery and delivery of bones to the Smithsonian marked a major breakthrough in the seven-year-old unsolved disappearance of two American journalists. "We are very hopeful that these are my brother's remains," Sam Blake told the *Times*. "What we are still not sure of is exactly what happened in those mountains before his death."

Soon after, Sam Blake called Owsley with answers to some of his questions.

According to a contact the Blakes had developed in Guatemala, Nick and Griffith were probably packing a tent at the time of their disappearance, since they were traveling in late March, the rainy season for the region. Both men were almost certainly wearing blue jeans—khakis too closely resemble military clothing and would have subjected them to potential danger. And they were probably carrying five to six days' supply of food in their packs.

Sam also reported that, as a result of a bad car accident in 1978, Nick had skull X rays on file at a Vermont medical center. But when he examined them, Owsley found no evidence of an irregular groove in Nick Blake's frontal sinus bone. So the two bone fragments did not belong to him.

On April 15, Owsley received a phone call from Colonel Al Cornell, the defense attaché with the U.S. embassy in Guatemala. After updating Cornell on the status of the forensic investigation, Owsley told him that he had found something unusual. Two of the bone fragments had metal fragments embedded in them. He was curious about weaponry used by the guerrillas.

Cornell explained that they had a variety of guns that discharged lead-core bullets with copper jackets.

After hanging up with Cornell, Owsley packed the metal fragments in an envelope and sent them to a firearms expert at the FBI laboratory for analysis.

Two days later, an Express Mail package arrived at Owsley's office from Griffith Davis's mother in Scranton, Pennsylvania. Owsley tore it open. It contained the skull X rays, including one of Davis's frontal sinus cavity. Immediately, he flipped on the X-ray viewing machine in his office and clamped the sinus X ray to the light board. It revealed an irregular groove in the sinus bone. This is him, he thought, thrilled at the first break in the case. It had to be. But Owsley could not be sure if he had that missing third fragment.

2

OPENING COFFINS

_____

April 30, 1992
Historic St. Mary's City, Maryland

White headstones dotted the ground of the Trinity Episcopal Church cemetery as Owsley, wearing a blue goose-down coat and a baseball cap against the spring chill, ducked under a tent erected to protect a large monument and the ground around it. A granite coping encased the ground beneath the tent, outlining a family burial plot. A historic preservation team of archaeologists and engineers that headed up "Project Lead Coffins: The Search for Maryland's Founders," welcomed Owsley. They had asked him to drive down from the Smithsonian to help them examine the contents of the only two seventeenth-century lead coffins known to exist in America.

While taking notes on his laptop, Owsley watched as the team dug back the soil and exposed a brick surface, through which they chipped a hole. Through the hole they dropped an extension ladder to the floor of an underground vault. It contained the coffins of Maryland's first royal governor, Sir Lionel Copley, and his wife, Lady Ann Copley. Born in England, the Copleys arrived in Maryland in 1691. Both died shortly after arriving, Lady Copley in 1692 and Lord Copley in 1693. They were the only seventeenth-century individuals known to have been buried on American soil in lead coffins, a sign of royalty or nobility.

But in 1990, archaeologists working at St. Mary's City, less than a mile from the Copley Vault, discovered the ruins of a 1667 Catholic Church, the first Catholic Church constructed in England's North American colonies. Maryland was founded in 1634, when Cecil Calvert, an English nobleman who had received a charter to the colony from King Charles I, established St. Mary's City as the colony's capital. The Calverts were Catholics, making Maryland the first English colony owned by a Catholic family. Calvert made religious toleration the official policy of the colony, and Jesuits built a church, later called Brick Chapel.

Catholicism was outlawed in Maryland in 1704, and the chapel was subsequently dismantled. The field on which it had stood was used to grow tobacco and wheat over the following two centuries, erasing all aboveground traces of the chapel. Eventually, Maryland declared St. Mary's City a living museum and authorized archaeologists to dig for the church's original foundation.

Aided by ground-penetrating radar, the archaeologists detected a dense mass concealed in a grave shaft. It was customary for seventeenth-century Catholics to bury their dead beneath church floors. The dense mass was probably lead coffins. Before exposing the coffins, the archaeologists wanted to learn more about them.

Since the only other lead coffins known to exist were the Copley coffins, the historic preservation team obtained permission from the vestry of Trinity Episcopal Church to examine them. The purpose was to study their construction in hopes of composing a plan to safely recover and preserve the ones believed to be beneath the foundation of the Catholic church. Owsley's role was to examine the Copley skeletons and provide an assessment of their condition. It would be a rare opportunity to look at remains from one of the oldest known burial sites in the American colonies.

The ladder firmly set, Owsley replaced his baseball cap with a yellow hard hat and descended into the damp and musty Copley vault. Reaching the dirt floor, he immediately spotted the lead coffins. Fallen bricks and broken mortar littered the ground around them. The tops of the coffins had been crudely cut off, the edges crimped and badly damaged. Owsley had expected that. Before his arrival, the research team had shared with him a historic letter dated August 1, 1799, written by Dr. Alexander McWilliams, a surgeon in the navy. When McWilliams was a medical student in the St. Mary's City area in the 1700s, he and roughly twenty colleagues broke into the Copley vault and forced open the coffins. Such actions under today's standards would have led to professional sanctions and criminal prosecutions. McWilliams's letter described what they saw when they pried open the coffins.

> We removed the lid and to our surprise saw within another coffin of wood. The lid of this being knocked off, we saw the winding sheet perfect and sound as was every other piece of garment. When the face of the corpse was uncovered it was ghastly indeed, it was the woman. Her face was perfect, as was the rest of the body but was black as the blackest Negro. Her eyes were sunk deep in her head. Her hair was short, platted and trimmed on the top of her head. Her dress was white muslin gown, with an apron which was loose in the body, and drawn at the bosom nearly as is now the fashion only not so low, with short sleeves and high gloves but much destroyed by time.
>
> In the coffin of the man was only the bones which were long and large. His head was sawed through the brain removed, and filled with embalmment, but he was not so well done as the other, or had been there much longer as he was much more gone.

Owsley knelt down beside the coffins. He lived for moments like this, the anticipation of what he would find beneath the lids fully captivating his attention. It was a rare chance, he felt, to climb into a time capsule and take a journey back to seventeenth-century colonial America.

Surrounded by archaeologists, Owsley was nonetheless alone in his thoughts as he opened the coffin. Lord Copley's eyes, both still preserved in a dry state, stared up at him. His short hair still existed on the front of his head.

At lightning speed, Owsley processed what he saw. All of Lord Copley's bones were in the correct anatomical position. The skeleton was in excellent condition, despite signs of some rainwater and debris that had undoubtedly fallen inside as a result of the original break-in.

Gingerly, Owsley lifted the two pieces of Lord Copley's skull and examined them. The cut mark was clean, indicating the skull had been sectioned consistent with seventeenth-century embalming procedures, a process that entailed removing the brain and organ tissue and filling the vault with spices and herbs.

Lady Copley's skeleton was in a more disarrayed state, with displaced bones. Looking at the skull, Owsley detected numerous tiny cuts, indicating the embalmer had made numerous false starts. The cuts were also cruder. Lady Copley's sternum had also been sectioned longitudinally.

For Owsley, the Copley skeletons were a book of knowledge, ready to be read and interpreted. They gave an insight into, among other things, seventeenth-century embalming procedures in colonial North America. In those days, opening the sternum to remove and replace vital organs with herbs and spices was common. But Owsley thought of another scenario to explain why the cuts on Lady Copley's bones differed from those on Lord Copley: an autopsy. He knew that autopsies were performed in the seventeenth century; in those days the scantily documented process involved sectioning the chests and craniums of the dead. Historical records from St. Mary's City confirmed autopsies being performed in the colony as early as 1642.

Since the vestry would not permit removal of the skeletons, Owsley handled each bone one by one, recording on his laptop their lengths, physical conditions, and any evidence of trauma. Then he documented the condition of the teeth from both skeletons.

When Owsley emerged from the vault, he assured the recovery team that he had gleaned enough information to prepare for the highly anticipated opening of the coffins beneath the Catholic church. The Copley burials would provide some context in which to compare the conditions of the other skeletons. Exhilarated, Owsley could hardly wait to return when it was time to enter the crypt beneath the old Catholic church.

3

BONE FRAGMENTS

May 13, 1992
Washington, D.C.

Owsley had completed his analysis of the materials shipped up from Guatemala, and was eager to discuss it with Randy Blake. That afternoon, Blake arrived at his office to learn the results.

"This is what we've done," Owsley said, leading him over to a countertop covered with ten-by-twelve-inch plastic freezer bags of soil. The bags were numbered 1 through 25. The bags contained all the soil from the two crates. He had X-rayed every bag. "These are some of the things we found in the soil," he said, pointing to tiny metal fragments, zipper teeth from blue jeans and one nail from a shoe or boot. "I've sent the metal fragments over to the FBI for analysis."

Owsley turned to his right. "Over here are all the bones, which we separated from the soil by screening with wire mesh. Now, there's not a lot of bone here."

Blake stared at distinctly colored sets of bone fragments; one set identified as "Individual A" and the other set as "Individual B." Individual A's bones were creamy yellow. Individual B's were white. Blake was intrigued by the difference in color.

"That happens with different bones of different people," Owsley said. "The burning process brings the color out. I've seen it in various situa-

31

tions." He explained that he had been to burn scenes and sorted out family members' remains partly by the bones' colors. The burning, combined with the bodies' differences in body fats and oils, and variations in how the minerals transform, can cause bones to color differently.

While all the fragments Owsley had collected represented only 3 percent or less of the bones from Davis and Blake, Owsley was positive there were two people represented here.

Owsley picked up two fingertip-size bones from the set marked "Individual B." One was a left malar, and the other was a right malar—cheekbones that support, respectively, the left and right eye orbit.

Then Owsley picked up another fingertip-size bone from the group marked "Individual A." "This is another left malar," he told Blake. "See, I'm looking for duplication of elements. I've got two left malars, telling me I've got two people here."

There were other duplicate bones. Owsley handed Blake two occipital bones, the skull bone at the back of the head. Not only did he have the base of two different skulls, but he also knew they belonged to males. Owsley explained that the occipital bone is where the powerful muscles that come up from the neck attach to the back of the skull. Those muscle-attachment sites tend to be more pronounced in males because their muscles are bigger and heavier.

Next Owsley picked up two other fragments with fine-looking fracture lines that zigzagged down the surface. "These are sutures, or the cracks that separate different bones in the skull," Owsley said as Blake looked on intently. "Growth occurs along those lines of union. If you took the suture down the middle of the skull of a child and put a nail on each side of that midpoint area, you would see those nails get farther and farther apart as the child grows, because that's where growth occurs."

In adults, the skull stops growing, causing the suture lines to start disappearing. "So you can start to identify a person's age by the suture lines on the skull," Owsley said.

Owsley pointed at one skull fragment; its suture was open. Then he pointed to another fragment; its suture was partially closed. "Do you see that?" Owsley said.

"Yes. There's a real difference in the two skull fragments," Blake said.

"This could be your younger guy," Owsley said, pointing to the open suture fragment. "And this more closed suture could be your older guy. This fragment is consistent with someone in his mid- to late thirties." Davis was thirty-eight when he disappeared. Nick Blake was twenty-seven.

"Griff Davis?" Blake asked.

"Well, I think so. Let me show you why else this might be him."

Owsley led Blake over to the X ray viewer. "Look at this frontal X ray of Griffith Davis's sinus. He has a very irregular sinus morphology."

Then Owsley put up the X ray of the two dime-size bone fragments that formed a partially reconstructed sinus. "If you notice here," he said, pointing to it, "there's an irregularity here too."

"I see it," Blake said.

"This is *probably* Griffith Davis. But I don't have enough to make a positive ID. And I can't identify Nick," Owsley said, pointing to the other fragments. "These *are* consistent with someone in his mid- to late twenties. But I can't definitely say it was your brother because I don't have enough evidence."

Owsley turned off the X-ray viewing light.

"What else can we do?" Blake asked.

"Well, we're on the right track here," Owsley said. And now he had gotten to the bottom line. He knew it meant more work and great risk, but it had to be done. "What we have is a piece of the story. But not the whole story. To really get this done right, we need to go there, locate the site, and dig it the way that it should be dug."

Owsley's offer to go to Guatemala and assist the Blakes in searching for their brother's remains delighted Blake. But it confounded him too. Owsley was an extremely busy man with no ties to the Blakes. Guatemala was far away, and the area they would have to search in was highly dangerous. Yet Owsley was volunteering to go. He had not even asked for payment or anything else in return for his services.

Blake didn't understand Owsley's motivation. It was not that Owsley *wanted* to go; he *had* to go. Owsley became a scientist because it fulfilled a desire to question and answer mysteries arising from skeletons. Many individuals—such as athletes, actors, models, and politicians—are extrinsically motivated, that is, they are motivated by external rewards such as money, praise, recognition, accolades, or a quest for power. Scientists often are intrinsically motivated; they are driven by the need to satisfy an inner urge to answer questions. But the really brilliant scientists go beyond high performance. They are driven to a point of compulsion. They become wedded to the science, as opposed to being merely a master of the science.

For Doug, the excitement of discovery takes over and he cannot rest or be satisfied until he has solved the mystery that the discovery presents. The boxes of soil brought up from Guatemala had ignited his investigatory juices, and his inability to solve the mystery was compelling him to go to Guatemala—no matter how great the risk—to dig for the answers.

"You think there are more bones in Guatemala?" Blake asked.

"There have to be more remains there," Owsley said. "There *have* to be. I'm sure of it."

4

GOING TO GUATEMALA

First week of June 1992
Museum of Natural History

WARNING. The word was stamped across the top of the State Department travel advisory Owsley had received.

"The Department of State warns all Americans to exercise caution when traveling in Guatemala," it read. "Violent crime is a serious problem, and armed robberies of public bus passengers are frequent. Visitors should avoid intercity road travel after sunset. Visitors driving cars should exercise particular caution; armed car thefts are common and those involved in car accidents are often put in jail." The advisory also suggested precautions to avoid cholera, yellow fever, and malaria.

Owsley's wife, Susie, had no idea he was going to Guatemala. The time had come to tell her—that evening, over dinner.

Susan Davies grew up four blocks from Doug in Lusk. Her parents—Marguerite Thomas and Marvin "Jake" Davies—met right after World War II in Denver, where Jake was stationed at the time and Marguerite was enrolled in a nursing program. Before Susie was born, they settled in Lusk, where Marguerite's family was from. Jake took a job setting up oil rigs in western states. He was often away from home, and Susie's mother worked as a secretary in Lusk, frequently leaving Susie in the care of her grandmother.

In second grade Susie went to Doug's birthday party. He decided then that she was the cutest girl in his class. He developed a crush on her before high school, so much so that by their sophomore year he assured her he would ask her to marry him. She hoped he would keep his word. Beyond his handsome face, Susie was much more interested in Doug's personality. His self-confidence, optimism, and down-to-earth approach to life made her want to be with him. She never saw him get discouraged.

After high school, she went to nursing school in Colorado while Doug attended the University of Wyoming. He immersed himself in his studies and stayed in contact with her. After college, true to his word, they returned to Lusk and married in the church located in their childhood neighborhood.

While he earned his Ph.D. at Tennessee, Susie worked as a nurse at the school's affiliated hospital and had two baby girls, Hilary and Kim. They were now ages fourteen and eleven.

"I'm going to Guatemala next week," he said, figuring it was best to just get it out of the way.

"Why are you doing *that*?" she asked.

He explained the situation.

"Doug, this sounds too dangerous, too risky." Susie leaned back in her chair, her eyes fixed on Doug.

"Oh, I'll be all right," Doug said. "The family knows George Bush, so they have some political connection. We'll be guarded by the U.S. military and the Guatemalan army."

"If you need to be guarded, then maybe it's not worth doing," she pressed.

Doug said nothing.

As a teenager, Susie fell in love with the way he treated her, as a gentleman who both communicated and showed his affection. And the qualities she saw in him—fierce loyalty and relentless persistence—convinced her early on that he was not just another boy who would be content to hang around Lusk for the rest of his life. No, he would develop into someone extraordinary. She had followed him to Tennessee, then Louisiana, and now D.C. At each stop, there was always something new to contend with.

In college, she tolerated the dead animals—roadkill, as she referred to it—that Doug would pick up from the highways and bring home in order to examine and study their bones. At Louisiana she gave up her summers with him so he could travel back to Tennessee and immerse himself in the bone lab there doing research. And since arriving at the Smithsonian, she had assumed virtually all of the day-to-day domestic responsibilities, from car maintenance to balancing the checkbook to child rearing, as Doug was increasingly traveling around the country— and sometimes the world—to work on cases.

She made the sacrifices because she knew and understood Doug fully. He was, to her, faithful, loyal, and reliable. He had no hobbies, played no leisure sports, and belonged to no clubs. He had no vanity and no vices; he did not drink, gamble, or smoke. He never raised his voice in anger or acted violently. He never spent money. All he did was work. But to Doug, his job was not work. It was an excellent and never-ending adventure. And Susie understood that. Doug was not consumed by a job. He was consumed by the pursuit of knowledge, and his job provided him with the chance to go after it. Susie was his safe harbor because she allowed him the space and the freedom to exercise his gift: this uncanny ability to see the past through skeletons. But when she felt he was neglecting his own health or welfare, she did not hesitate to intervene. It happened once while they were at Tennessee.

The year was 1981, and Doug had decided to spend the summer doing research at Tennessee. They brought the girls—then three and one—and lived on campus in Knoxville for the summer in order to be near Doug. One day Bill Bass stopped in the bone lab to talk with Doug and found him sitting down, fatigued and short of breath. Bass had never seen Doug tired, even when he used to study virtually around the clock. He encouraged Doug to go see a doctor at once. Instead, Doug went home.

That night, he started coughing up blood. Susie did not hesitate. She brought him to the university hospital, where she used to work as a nurse. Left to his own devices, Doug would not have sought medical attention. Susie set him up with a doctor she used to work with, explained Doug's symptoms, and predicted he had a lung infection of sorts.

Doug provided sputum specimens—spit elicited from the lungs— and was told the results would be back within a week. In the interim, he went back to work. Then the doctor summoned him back to the hospital.

"I'm not going to beat around the bush," he told Doug. "I'm just going to tell you. I think you have lung cancer." He wanted Doug admitted within twenty-four hours to begin more advanced testing. Doug's first thought was: I have to beat this. I have to fight it.

But as the news sank in, so did the realization: There was no cure for lung cancer.

The only place he knew to go was Susie, his safe harbor. He drove straight home. The moment he walked through the door, she knew something had gone terribly wrong.

"What did the doctor say?" she asked.

"Well," he said slowly, "I have lung cancer."

Face-to-face, they stared at each other in silence. Doug handed her the test results. "Oat cell carcinoma," the report said. "Positive for malignancy." In her days at the hospital she had read many similar diagnoses, and those words were a death sentence. She folded the paper, her eyes welling up.

She looked up at Doug, subdued and looking lost in his own home. That was the first time she had ever seen Doug vulnerable. A tear trickled down Susie's face as she threw her arms around him and pressed her lips to his cheek. He had just turned thirty. "I'm not going to go down easy," he whispered unconvincingly.

Later, Doug stood in the bathroom doorway and watched silently as Hilary and Kim played innocently in the bathtub. Bent over the tub, Susie fought back tears, struggling to soap them. They are going to be without a father, she thought.

By the time she had dressed them in their pajamas, she had decided to start acting like Doug's nurse instead of his wife. And she applied his approach to problem solving: attack it immediately and relentlessly. Before going to bed, she called the doctor at home and challenged his diagnosis, pointing out that Doug's age and habits did not fit the profile of a lung cancer patient.

The doctor stood his ground but offered to take another sputum specimen. She said that was a waste of time and insisted on a bronchoscopy at once. The following morning she took Doug to the hospital, where a scope was inserted down his trachea to explore his lungs. He remained hospitalized for forty-eight hours. The results confirmed that there were no tumors on his lungs. Rather, he had scar tissue and inflammation, a condition that can make normal cells look different and appear like cancer cells. Further consultation with the doctors traced the source of Doug's lung infection to the bone lab. Being underneath the university's football stadium, the lab was unusually damp and musty. Worse still, many of the bones Doug was working on were mildewed, and he was handling them for up to sixteen hours per day. The high exposure had likely weakened his lungs, making them susceptible to infection.

Susie had shepherded Doug through a medical crisis. As soon as he received antibiotics to treat the infection, Doug returned to the bone lab. Bass brought in some dehumidifiers. She tolerated that. But charging off to Guatemala was different. The stakes were much higher.

"Doug, it's too dangerous," Susie said, picking at her food, then putting down her fork. "Americans are killed every day in situations like this. They wouldn't think twice about killing you too."

Doug kept silent and stared at his food.

"Doug, you've got responsibilities," Susie said, getting up from the table and walking into the kitchen. "You've got kids. It's too dangerous to go."

Doug still said nothing as he began to eat. The pull to go was irresistible. The dangers only made it more enticing, more adventurous. Cases like this were what had appealed to him when he applied to the Smithsonian.

Susie returned to the table and said nothing more. When he was like this she knew he wouldn't change his mind. She vowed to herself not to tell the girls where their father was going. And she prayed he would return.

5

OUTSMARTING THE DEVIL

June 11, 1992
Nebaj, Guatemala

Flying at an altitude of 8,600 feet, the twin-engine plane veered danger-ously close to a mountain as it dipped into a landing pattern. Seated beside Owsley and looking out the window, fellow anthropologist John Verano saw nothing but a vast green mountain range that stretched as far as the eye could see. Fluent in Spanish and having worked at the Smithsonian since 1987, Verano had been invited by Owsley to assist him.

"Looks like we're here," said Colonel Al Cornell, the military attaché from the U.S. embassy. He was all business with his blue military cap, khaki uniform, and a Swiss 9-mm pistol.

Owsley and Verano looked down, seeing only a tiny gravel road with some military bunkers beside it. "Where's the runway?" Owsley asked.

"You're looking at it," the pilot said.

Owsley removed his navy blue baseball cap with the words *Smithsonian Institution Physical Anthropology* emblazoned across it in yellow stitch and scratched his head. It wasn't what he had expected.

The plane touched down, its wheels churning up gravel and mud as it skidded to a stop beside a convoy of vehicles: two Nissan Pathfinders, an Isuzu Trooper, a Suzuki Samurai, a white Nissan double-cab pickup truck, and a blue U.S. embassy Suburban.

Owsley grabbed his gray nylon carry-on bag and his archaeological screen—a lidless wooden box with a wire mesh bottom for sifting dirt and separating out bone fragments. Verano tossed out his backpack and camera bag. Randy and Sam Blake, who had rented the plane and the vehicles, began loading shovels and bags into the Suburban.

"Dr. Owsley, why don't you and Dr. Verano ride with me?" said Guatemalan army lieutenant colonel Otto Noack, who was an imposing six feet, three inches tall. Middle-aged and soft-spoken, he would be their military escort. Unlike the American colonel, he was dressed casually in a white, short-sleeve polo shirt and blue jeans. "We'll be going in the Pathfinder."

"OK," Doug said, tossing his belongings behind the backseat.

Colonel Cornell climbed into the Trooper. The Blakes' guide-for-hire, civil patrol leader Felipe Alva, slipped into the pickup truck with another member of the army's civil security force. The Blakes still owed Alva eight thousand of the ten thousand dollars they had offered to pay him for delivering their brother's human remains. He would not get the balance until leading them to the burn site.

After an hour of driving, Alva stopped the convoy at a dirt path, telling everyone that the burn site was only reachable on foot. Following a two-hour hike over steep hills and rugged terrain, Alva finally stopped.

"*Alli,*" Alva said, pointing down into a steep bowl-shaped valley. "*Este es el lugar donde los restos humanos fueron quemados.*"

"What did he say, John?" Owsley asked his colleague between breaths. They were all breathing hard from the exertion.

"He says that the remains were burned down there," Verano said.

The Blake brothers were angry at the length of the hike and skeptical of Alva. Squinting, they strained to see precisely where Alva was pointing. A vast corn and potato field blanketed the landscape beneath them, a misty fog hovering above it. The path leading to the field wound through a deforested area. Some of the land had been logged, while other stretches of forest had been destroyed simply to reduce cover for the guerrillas.

Suddenly Alva sat down on the edge of the path. "I'm going to sit here out of respect for the dead. He will lead you to it," Alva said, pointing to a short Guatemalan man in his thirties who had been alongside Alva since the start of the trek. Wearing a white cowboy hat, a black nylon jacket, and brown dungarees, with a machete attached to his belt, the man started briskly down the path without saying a word.

Owsley and Verano looked at each other.

"I'll stay with him," Verano said, readjusting his pack as he hustled ahead of the others.

Passing uprooted tree stumps and stepping over dead tree boughs that obstructed the path, Owsley and the others lugged tools and bags until they reached the cornfield. The corn stalks brushed just below the pens and pencils clipped to the chest pocket on Owsley's denim shirt. Finally, Alva's assistant stopped.

"*Este es el lugar,*" he said.

Verano turned to Owsley and the others. "He says this is the place."

While the others caught their breath, Owsley put down his bag and assessed the area. He noticed a thin layer of ash and charcoal on the brown soil. His first thought was that the burn area was too small. Then he noticed another inconsistency: there were not enough trees around the immediate vicinity to build a big fire.

Skeptical, he knelt with a small trowel and began peeling back soil. The others formed a semicircle around the area. Verano screened the soil that Owsley dug up. The Blakes hoped they'd finally have their answers.

No one spoke as Owsley felt the soil through his fingerless black gloves. Then he picked up a handful of it. Brown and moist, it did not match the red, claylike soil contained in the crates that Alva had previously delivered to the Blakes for analysis at the Smithsonian. Nor did the plant roots in the soil match the roots found in the crates.

Verano helped Owsley shovel dirt into the screen. Shaking it to separate the soil from any artifacts, Owsley found no bone fragments. He stopped working and looked at Verano, who was equally puzzled. "This guy's lying to us," Owsley said softly.

Both men stood up and brushed the dirt off their knees. Owsley turned to Sam and Randy. "This isn't the site," he told them.

"Damn it. F—— this," Sam seethed, clutching the straps of his 35-mm Nikon around his neck.

"Are you sure?" Randy asked.

"There's not enough ash and charcoal present to account for the extensive burning in the bones," Owsley said. "There's not enough wood around to build a fire. And the soil and roots here don't match what came to my office in the coffins."

Sam cursed and removed his glasses.

"Confront Alva's assistant," Randy said, looking directly at Verano. "Tell him what you've concluded."

Verano, flanked by Colonel Cornell and Lieutenant Colonel Otto Noack, approached Alva's assistant.

"Sí, este es el lugar," the assistant said, fidgeting. *"Vine aquí en marzo y cave la tierra yo mismo, una area de cincuenta pies cuadrados."*

"What's he saying?" Sam asked.

"He says this is the site," Verano said. "That he came here himself in March and dug here."

"It can't be," Owsley said, his hands on his hips. "It's not right."

Bringing his right index finger and thumb together to form a point, Verano moved his hand back and forth to emphasize his words. *"Hemos examinado este sitio y no puede ser el lugar donde los restos fueron quemados y enterrados,"* Verano told the assistant.

The assistant hesitated. *"Estuve aquí. Este es el lugar,"* he began, motioning with his arms as he went into a long explanation.

When the assistant finished, Verano turned back to Owsley. "He said, 'I was here. There were men with guns on the hill. I was afraid. I was digging with a machete. I dug out as much as I could. I filled it up. And I got out of here.'"

All eyes turned to Owsley, who let out a deep breath. "He's lying," Owsley said.

His arms still folded, Noack took a step closer to Alva's assistant and stared down at the smaller man. *"Que pasa aquí?"* Noack asked.

The assistant reassured Noack that he had led them to the true burn site.

"Pues, los doctores no te creen," Noack said. *"Eso no lo creen."*

"We're being set up here," Sam Blake said.

"Doug, do you *really* believe he's lying?" Randy asked. "Do you really believe this isn't the right place?"

"There's no question in my mind."

Randy threw his hands up in the air. "This is a joke," he complained. "Alva misled us."

The assistant suddenly bent down and reached behind a burned log on the ground. He retrieved a very small green plastic bag and handed it to Noack, who passed it to Owsley. Alva's assistant told Noack that he had left the bag under the log months earlier when he was excavating the site and filling up the two crates that Alva eventually gave to the Blakes.

Owsley looked inside the bag. It contained nearly twenty coin-size bone fragments. "They're definitely human," Owsley said, passing a couple to Verano to inspect.

Craning their necks, the rest of the group crowded around Doug.

While Verano looked at the bones, Owsley closely examined the soil and the bag. The bag weighed about eight ounces and was in good condition. It showed no signs of weathering. Yet it had supposedly been out in the field for a couple of months. While the bag had almost twenty small bone fragments in it, there were no bones on the ground. And the soil in the bag was a red clay type, just like what had been delivered in the coffins. Yet there was no red clay in that area.

"Doug, what do you think?" Randy asked.

Owsley looked up at Noack and Cornell. "The soil in this bag didn't come from this site," Owsley said. "The soil doesn't match."

"What does this mean?" Sam asked.

"The fifteen or twenty tiny human bone fragments in this bag were probably removed from the actual burn site and brought here," Owsley said. "I think this guy may have wanted to get down here before us and sprinkle that little bag with the bones and ash around the site, to make it *look like* the real site. But we kept up with him and he didn't have an opportunity to dump the bag out on the ground."

Everyone was convinced. Furious, the Blakes decided it was time to speak to Alva.

Thirty minutes later, they found him. Noack expressed Owsley's conclusion to Alva, who shrugged his shoulders and spoke back to Noack in Spanish.

"He says that he doesn't know the place," said Noack in perfect

English as he turned to Owsley and the Blakes. "And that he's separate from the civil patrollers who know the exact location, and that he never actually saw the place but that his people told him where it was."

The Blakes were incredulous. After huddling to discuss their options, they confronted Alva.

"You took us to a false site," Randy shouted.

"Me dijeron que este fue el sitio," Alva said.

"Look, we're not leaving this place until you take us to the real site," Randy said.

"We've got to go to El Llano," Sam said. "That's where my brother spent the night before he was murdered."

Reaching the village of El Llano required passage through a twenty-mile dangerous stretch, patrolled by roaming bands of armed paramilitary forces. Although Noack held military rank over the paramilitary bands, he knew that taking Owsley and the Blakes there would require additional military support and helicopters. The Blakes offered to pay for the chopper rentals if Noack supplied the military support.

"We don't care how you do it," Randy said to Alva. "But we want you to take us there."

Alva scowled. *"Me voy a casa,"* Alva said in defiance. *"Voy a manejar a Huehuetenango."*

"He says he's going home to Huehuetenango," Noack said.

"You double-crossed us," Sam shouted. "We don't wanna have this happen again. We're down here in good faith. And you're not gonna get another dime from us—which is what this is all about for you—until we get performance. Take us to the remains."

Noack stepped directly into Alva's face and began speaking in rapid Spanish.

"What's he saying?" Sam asked Colonel Cornell.

"That this is a direct order from the army high command. This is a high-priority case. And you *are* going to help us. Basically, he said he'd kill him if he doesn't help."

Alva paused. *"Está bien. Está bien. Iremos al Llano."*

June 12

El Llano

"Here, Doug, you may need this," Lieutenant Colonel Noack said, handing him a grenade.

Owsley stared at it.

"Put it in your pocket," Noack said.

Clutching the grenade tightly, Owsley slipped his closed fist into his pants pocket. He had no idea how to use it but assumed he would figure it out if he had to. He followed Noack, today dressed in military fatigues and fully armed, and a half-dozen Guatemalan army rangers he had recruited to accompany them into the village. The rangers guarded Owsley and the rest of the Blakes' search party as they and Alva canvassed El Llano in search of people who had information on the burn site location. Eventually, the group found the schoolhouse where Griff Davis and Nicholas Blake spent their last night alive. Through bits and pieces of intelligence, the group concluded that Davis and Blake were burned and buried outside El Llano in a mountainous area more accessible by helicopter. The Blakes agreed to reserve choppers for the next day. Alva said he thought he could get someone in the village to tell him the precise burial location. They didn't put much stock in his word, but they had little choice.

June 13

About ninety miles northwest of Guatemala City, the two helicopters flew around a mountain and came upon a small village of huts. Like ants scrambling from a dirt hill that has just been kicked, dozens of men, frightened by the sight of Americans and the chopping sounds of the helicopter blades, tore out of the huts with their guns and sprinted into the jungle.

After touching down, Owsley and the Blake brothers waited while Alva and Noack searched in a nearby area for the burn site. An hour later they returned.

Noack was upset and told the Blakes he would need to borrow a helicopter and would return as soon as possible. He left them his Uzi and some grenades.

Thirty minutes later, Noack returned with a member of Alva's civil patrol unit. "We know where it is now," Noack shouted from the helicopter, which was hovering above the group. "It's about a one-mile hike from here. The pilot says we've got about three hours. Then the rains will be here."

Eager to go, Owsley looked at the Blakes.

"All I know is that we may never get back to this place," Sam said. "We're here now. Let's do it."

Noack took his Uzi back and strapped it over his shoulder. With Alva's man leading the way and Noack guarding the rear, the group hiked until they reached an area where the ground was littered with charred logs. His plaid flannel shirt stained with sweat, Owsley pulled a spade from his bag. Bending on one knee, he made rectangular cuts in the soil, sectioning it off into quadrangles. Then he carefully began peeling the soil back with a trowel. Immediately, he saw human bone fragments. He picked up a handful of soil and looked at it closely. The red clay texture matched the soil shipped to his office.

"This is the place," Owsley said, starting to dig quickly.

"Here," he said to Sam and Randy, handing them trowels. "Start digging in this area. Just work the soil like this." Doug inserted his spade a couple of inches into the ground, then pulled it back, probing for fragments. "Oh yeah, see this." His adrenaline racing, he held up a small fragment. The trip was going to be a success after all. To have failed after all the risk would have been unacceptable.

Owsley marked off a wider area that ran alongside a series of charred logs. Sam started digging. Randy spread a black plastic bag on the ground and placed the screen on top of it. Doug and Sam began dumping soil into the screen. As Randy shook the box-shaped screen, soil fell through to the bag while bone fragments surfaced on the wire mesh.

"Look at this," Sam said. "A burnt tent stake."

"Yes," Doug said, taking it from his hand. "And it matches the stakes that came up in the coffins. This is definitely the place."

Suddenly Sam froze. The tip of his spade had hit a tiny object. It was a wire rim with half of a melted eyeglass lens. The lens had a yellowish tint. Sam picked it up and studied it. Nick Blake had had a pair of gold wire-rim glasses. And he had been a heavy smoker, accounting for the slight discoloration of the lens. Due to a smoking habit, one of the lenses had a unique yellowish tinge to it. "My God," Sam whispered. "These are Nick's glasses."

It suddenly struck Sam that he was excavating his brother. Randy and Doug both stopped and looked at Sam, whose hands were shaking and whose eyes started to well up. Randy took the lens from his open hand. "My God, Sam. We've really found him."

Goose bumps surfaced on Owsley's arms. Silent, Otto Noack stared down at the lens and wire rim from his perch on a burnt log.

"That's very important, Sam," Owsley said softly. "Put that in a separate bag."

Owsley continued to cut away the soil, widening the excavated area to over four feet in diameter.

Struggling to control his emotions, Sam picked up his spade and returned to work.

"Oh, this is good stuff," Doug called out, his voice getting louder and more enthusiastic. "We're moving into the fire area. We're hitting the main area."

Sam turned to look at Owsley, whose hands were fully immersed in the soil. Beads of sweat ran down his face, and sweat stains had formed along the edge of his baseball cap. Completely engrossed in unearthing the remains, he was impervious to everything around him—the weather, the danger, or the fact that everyone was watching him. Even the high elevation, which forced the others to labor for breath, did not seem to affect Owsley. At that moment, Sam was grateful to have Doug there. While some might find his enthusiasm strange, to Sam it was a sign of his passion and caring.

"Oh, look here," Owsley said. He pointed down to a cluster of eight white fragments slightly smaller than miniature marshmallows. Removing his soil-stained gloves, Owsley delicately picked through tiny white roots and green plant leaves, lifting up the objects. "Teeth," he observed. Two of them contained the crown portion. None of them contained fillings.

Owsley remembered that most of Davis's teeth had fillings in them. None of Blake's did. "This is very important," he said. "These teeth may belong to Nick."

"What do you think of this?" Sam said, handing Owsley a bone fragment.

"It looks like a piece of a jawbone," he said. "Very good. Very important."

Randy continued to shake the screen, separating bone fragments from soil, then placing the bones in plastic bags.

Owsley again removed his tape measure and extended it over the excavation. They had exposed a five-foot-by-five-foot area.

Overhead, darkening clouds started moving in rapidly. Owsley looked at his watch. They had already been there almost two hours. Noack emerged from the bushes. "The chopper pilots are saying we have to get out of here," he shouted, his voice competing against the escalating wind. "Bad weather's coming."

"OK," Doug yelled up, anxious to screen all the soil they had uncovered.

"Wait a minute," Sam yelled. "Do we have enough? Do we have enough?"

Owsley looked at the fire hearth. "Here's what we're going to do," he said quickly. "We're going to dig up all of the remaining ashes that we haven't screened. Get some of those big garbage bags."

The Blakes grabbed some ten-gallon plastic bags and opened them. Owsley scooped chunks of soil and ash with his trowel and dumped it into the bags. He quickly dropped the trowel, cupped his hands together, and began scooping the soil with his hands. Over thirty minutes, he filled four bags.

"OK," Doug said, breathing hard. "We've got enough."

"Doug, you're sure?" Randy asked. "Because we'll stay up here overnight if we have to in order to finish this."

"We've got enough."

Everyone in the group scurried around the hole, picking up all of the Ziploc bags full of remains and artifacts. Others hefted the larger black garbage bags.

"Let's go," Noack called out.

In single formation, they followed Noack down the trail toward the sound of helicopter blades cutting through the air in the distance.

After they loaded everything and climbed in, the realization of what they had just done—excavated the grave of two cremated Americans—set in on the entire group. No one said a word as the choppers eased upward, then sped over the trees, away from the storm.

Sam Blake removed his glasses, rubbed his eyes, and buried his head.

As always, Susie was waiting at the airport when Doug arrived back from Guatemala. She was terribly relieved to see him get off the plane. But her emotions were mixed. She had still not forgiven him for going. She had worried for his safety the entire time he was gone and had hoped that no news was good news. Because of the remote region he was in, Doug had not been able to call her. To Susie, he had unnecessarily put himself in harm's way to satisfy a selfish desire for adventure.

The car ride home was awkward. Susie almost never saw selfishness in Doug, and she rarely got angry at him. This was an exception, and Doug knew it. He also felt he deserved the cold shoulder he was getting. Much like the time his parents caught him with a dissected frog, Doug figured he had to endure some justified heat from Susie. So he eased gingerly into a summary of what happened. He purposely left out any mention of grenades, machine guns, or his confrontation with the guide whom he had accused of lying about the burial site.

June 17
Washington, D.C.

When Owsley arrived at the Smithsonian, two parcels awaited him. One was a letter postmarked from Cambridge, Massachusetts. The other was a Federal Express package from Philadelphia. Doug opened the letter first.

"Dear Doug," it began. "Enclosed is a picture of Nick with his wire-rim gold glasses on. They are probably the same ones we found at the site.

"The photo is from August 1984, six months before he was lost," the letter continued. "Good luck. Best, Sam Blake."

Owsley opened the Federal Express package. It contained a hand-written note from Thomas Cush, Nick Blake's dentist.

"I sent all the X rays we have and his original work-up sheet," Dr. Cush's cover letter stated. "We did not do the root canal on #8. There was a crown and post made for #8 as Nick broke it the summer of 1968."

Owsley retrieved a black X-ray sheet bearing Nick's name and compared it to a postage-stamp-size fragment of jawbone that Owsley had recovered from the burn site in Guatemala. They matched. So did the wire-rim glasses from the dig with the ones Nick was wearing in the family photograph.

As for Griff Davis, the irregular groove that Doug had previously detected in the two fragments of his sinus cavity proved critical. While sifting through the hundreds of tiny bone fragments he recovered from the burn site, Doug found the missing third fragment of the sinus cavity. Now complete, the sinus cavity perfectly matched the X ray of Davis's sinus cavity.

It all confirmed Owsley's conclusions. Despite overwhelming odds, his persistence had resulted in certainty and closure for the Blakes and Mrs. Davis.

Over the next few days, he prepared his report, then contacted the U.S. embassy in Guatemala with the news. Nick Blake and Griffith Davis were murdered, then buried in a wooded area outside a remote village in Guatemala. A year later their bones were excavated, transferred to a new location, crushed into small pieces, and burned. The embassy immediately sent out a telegram to U.S. Secretary of State James Baker:

> Smithsonian medical anthropologist Dr. Douglas Owsley called Embassy June 22 to report he had been able positively to identify the remains of Nicholas Blake, based on three right molars of the lower jaw which were recovered in mid-June expedition reported. Given the extensive White House and NSC interest in this 7-year-old disappearance case, the Ambassador is writing a letter to President Bush informing him of the news.

Case No. 92-3 was closed.

6

INTO THE CRYPT

November 9, 1992
Chapel Field
Historic St. Mary's City, Maryland

Ever since laying eyes on Sir Lionel Copley and Lady Ann Copley in the lead coffins buried beneath Trinity Episcopal Church, Owsley had conjured up images in his mind of what he'd find in the lead coffins entombed in the foundation of seventeenth-century Brick Chapel, which shared the same property. The day had finally arrived to look inside.

The sign reading LEAD COFFINS had an arrow pointing toward a ten-acre field encased by an old farmhouse, a barn, and tall trees. In the middle of the field, television satellite trucks parked along the outside perimeter of a chain-link fence erected around the exposed foundation of the Brick Chapel. After showing his credentials to a security guard at an opening in the fence, Owsley entered, walking past two portable toilets and into a green-and-white nylon tent that covered the crypt found beneath the church's foundation. Under the tent, the U.S. Navy had constructed a cradle system to hoist the lead coffins—two of which weighed five hundred pounds or more—from the crypt. Before their removal, an engineer from the Armed Forces Radiobiology Research Institute used nuclear-powered gamma rays to take X ray–like pictures through the metal and discovered that only one coffin appeared sealed.

NASA scientists then used a tiny borescope to extract an air sample from the sealed coffin. They hoped to obtain seventeenth-century air that could provide priceless information about changes in the earth's atmosphere over three centuries. But even the ancient air taken from the sealed coffin contained some modern gases composed of man-made substances that did not exist in the 1600s, including Freon.

While waiting for the first coffin to be lifted from the crypt, Owsley left the recovery tent and entered an adjoining tan field hospital tent set up by the U.S. Army Reserve. It was equipped with examination tables and medical instruments, and provided a controlled environment in which Owsley and his colleagues could remove and analyze the contents of the coffins. While Owsley put on a white medical coat and rubber gloves, his assistants, twenty-six-year-old Kari Sandness and twenty-five-year-old Pam Stone, prepared for the first coffin to arrive.

Sandness laid out the magnifying glasses, tweezers, brushes, dental picks, rulers, and other instruments needed to analyze the skeletons. Owsley had met Sandness while visiting the University of Nebraska, and after she obtained her Master's degree in physical anthropology, Owsley hired her as his personal assistant. Behind her Meg Ryan looks, Sandness had an unusually high level of skill and professionalism for someone fresh out of graduate school. Her job title was officially Museum Specialist. Unofficially, she was Owsley's troubleshooter.

Stone, a new intern who was about to start her graduate studies in anthropology, set up a laptop. Her internship duties included making an on-site computerized record of all the skeletons Owsley examined. Smithsonian photographer Chip Clark, a burly, barrel-chested man, rounded out the team. He positioned his equipment alongside the main observation table. His photographs would become part of the historical record archived at the Smithsonian.

As soon as they were ready, the first coffin, a crudely made lead box, was raised up and brought before Owsley. It was less than three feet long and eight inches wide. Inside they found a wooden box, the top of which had deteriorated. Soil was inside it. Beneath the soil, Owsley spotted the skeleton of an infant. Its lower legs were missing; some shroud pins and finely woven linen remained.

Owsley looked at the skull and noticed a cluster of tiny circular holes. With a flashlight, he leaned over the coffin, peering directly into the holes. He searched for fracture lines or other signs of trauma on the skull. There were none, leaving him to conclude that the holes had been caused by something internal, a cranial infection of some sort.

Carefully inspecting the infant's other bones, Owsley saw lesions on the ribs, a sign that, when coupled with the skull's condition, convinced him that the infant had had rickets, scurvy, and chronic anemia. He had seen the same symptoms in some American Indian infant skeletons racked with rickets, a disease that stems from a lack of vitamin D and insufficient exposure to sunlight, resulting in defective bone growth. The holes in the infant's skull were likely the result of an inflammation tied to the rickets. He figured the Colonial practice of bleeding diseased patients had probably exacerbated the infant's anemia. "This was a terribly sick little child," Owsley said.

Using tweezers, he collected tiny bone fragments and had them taken to the portable X-ray tent. Stone recorded a complete inventory of each bone.

Next the engineers brought in a five-hundred-pound coffin. It too contained a wooden box, with an adult female skeleton. She had been buried in a linen shroud. Her wrists, knees, and feet were bound in silk ribbon. Rosemary sprigs lay across her body. Everything was consistent with seventeenth-century English burial practices.

Instantly it was clear to Owsley that he was witnessing the best-preserved seventeenth-century colonist ever discovered in North America. Desiccated skin tissue still adhered to the bones. Body and scalp hair were present. Most of her teeth were missing, but the ones that remained showed signs of abscess and extremely heavy wear.

Examining her bones, he noticed that her right leg was shorter than her left leg. The right femur had a spiral fracture. Judging by the condition of the bone regeneration along the fracture lines, he gauged that the woman had broken her leg two to five years before she died and that it had been improperly reset, resulting in her walking with a limp until she died. The poorly set leg might have had some connection to her death too, as the bones from the lower part of her right leg were inflamed, indicating she was suffering from infection at the time she died.

As the intrigue and mystery over the identity of the skeletons mounted, the number of reporters and interested spectators from the surrounding county grew outside the medical tent. By the time the final lead coffin was hoisted from the crypt and brought into the examination area, ABC's *Nightline* crew was on hand to observe. The coffin was the largest, best-crafted of the three, a six-sided box with iron handles at the head and foot. A cedar coffin rested within.

When the engineers opened the lid, Owsley was speechless. From the waist down, the skeleton's bones were in excellent condition. But the upper body and nearly the entire skull were reduced to white crystals. Almost no bone remained. Yet eight-inch strands of shoulder-length auburn hair were perfectly preserved.

Examining the hip and leg bones, Owsley could tell the remains were male. Measurements confirmed he was about five and a half feet tall at the time of his death. None of his bones showed signs of trauma or infection.

Speculation over the skeletons' identity mounted. Surely they had been Catholic. And the lead coffins and elaborate burial confirmed that they had been wealthy and most certainly of nobility. Plus, they had to have died before the Brick Chapel was dismantled at the close of the seventeenth century.

The next step was to transport the skeletons to Owsley's lab at the Smithsonian so that he and his team could confirm their age, sex, race, and cause of death. Historians were busy checking St. Mary's City records for Catholics who died prior to 1700 and would have qualified for lead-coffin burials. Pollen samples were taken from the coffins, the results of which would determine the season that each individual died. Scientists also collected over a dozen types of insects, which constituted the only substantial collection of seventeenth-century American insects. Once identified, they would help confirm the seasonal evidence provided by the pollen samples. And hair samples were collected from both adult skeletons, another seventeenth-century first. Since hair generates a record of ingested foods as it grows, analysis of hair strands can provide a record of diet in the months leading up to a person's death. With all this information at his disposal, Owsley's task was to help answer the ultimate question: Who were the individuals in the coffins?

December 2, 1992
Smithsonian Museum of Natural History

Snow fell lightly as a black hearse backed up to the rear loading dock and a tall, well-built funeral-home director in a dark blue trench coat stepped out. Owsley, in his white lab coat, helped him wheel out a contemporary silver casket with chrome handles and maneuver it onto a yellow hydraulic lift. Elevating to the dock, they wheeled the casket into the museum and up to Owsley's office.

Owsley's assistant Kari Sandness removed clear plastic containers that held the remains recovered from St. Mary's City. One by one she unpacked the bags and assembled them on the table in anatomical order. While she documented, Owsley identified the remains.

His analysis of the adult male skeleton went quickly. Without a cranial vault, Owsley couldn't certify the man's race. Nonetheless, he knew the individual had been Caucasian, as few Blacks were present in early colonial Maryland and it was most unlikely that anyone other than a Caucasian would receive such a royal burial at that time. Owsley detected arthritis in the individual's knees, ankles, and various other bones. Analyzing the arthritic bones under a microscope and the seventeen teeth, all of which showed almost no wear or calculus deposits, he estimated the age at death was in the range of forty-five to fifty-five years.

So the profile was complete. Coffin one contained a forty-five- to fifty-five-year-old white male who stood five feet, six inches tall, with shoulder-length reddish hair. His bones showed no signs of trauma or infectious disease. At the time of his death, he was suffering early stages of arthritis.

When Owsley lifted the bones of the adult female, he sensed they were unusually light. The more he handled them, the more convinced he became that she had had severe osteoporosis, a disease that erodes bone density. The bones also showed signs of widespread arthritis and malnourishment.

Using a dental pick, he examined her remaining teeth. Twenty of them had abscessed or been lost prior to death. None of her remaining teeth were touching each other. Observation of her hair samples showed that she had ingested large doses of arsenic in the weeks leading up to her death. The arsenic in her system could have been a result of poisoning. But Owsley doubted it. The condition of her bones suggested the arsenic was probably administered as a medication to treat her disease.

After taking measurements, he estimated her age at fifty-five to sixty-five years at the time of death. She was five feet, three inches tall and walked with a limp due to her leg fracture.

The third coffin contained a five- to six-month-old infant, likely a female that was suffering from rickets and iron and vitamin deficiency at the time of her death.

Results from the pollen lab suggested the infant was buried in the spring.

The adult female's coffin had signs of ragweed, indicating an autumn burial, while the lack of pollen in the male's coffin suggested a winter burial. The insects from all coffins confirmed the findings.

Owsley's composite information was turned over to the research recovery team and coupled with the other profile data gathered by researchers and scientists. Only one man matched all the evidence: Philip Calvert, the youngest brother of the colony's founder and a man of wealth and power. The half brother of Lord Baltimore, Philip arrived in Maryland in 1657 and served as governor from 1660 to 1661. From 1657 to 1682 he was the colony's chief judge, negotiating and enforcing all treaties with American Indians. He also restored religious toleration. In 1678 he built a large home in St. Mary's City and called it St. Peter's. He owned 3,900 acres when he died in 1682. William Penn was among those who paid their respects.

A process of elimination identified the female skeleton. The coffin beside Philip Calvert's was positioned in a manner typical of the way husband and wife would have been buried in those days. Philip's first wife, Anne Wolsey Calvert, fit the age profile Owsley found. Born into a prestigious English Catholic family, Wolsey married Philip in England and died in Maryland in 1680.

Identifying the Calverts deepened the mystery of the child's identity. Historical documents said Philip died without children. And Owsley's age profile of Lady Calvert placed her at approximately sixty when she died, making it virtually impossible that she could have given birth to the child. Nor would her deteriorated physical condition been conducive to child bearing.

Also, infants were rarely buried in lead coffins. Even royal children in England usually did not receive lead-coffin burials, especially female children.

The focus next shifted to Philip's second wife, Jane Sewall, a younger woman whom he married shortly after his first wife's death and just before his own passing. Researchers agreed that the baby was probably hers. After Philip's passing, Sewall returned to England, where she later died and was laid to rest.

At the conclusion of Owsley's work, all three skeletons were stored in his lab. Eventually, they would be returned to historic St. Mary's City for reburial underneath a yet-to-be-constructed new Brick Chapel, complete with historical markers indicating their identity and honoring their contribution to Maryland's colonial history.

7

BUILDING PEOPLE

April 19, 1993
Zagreb, Croatia

Exhausted, Owsley flipped on the television in his hotel room and sat on the edge of his bed. For the past week, he had been touring medical facilities and universities throughout war-torn Croatia, lecturing and assessing the scientific community's inability to handle the country's new, large-scale human rights dilemma. The recent breakup of Yugoslavia and the ensuing Croatian civil war had left the tiny country with fourteen thousand war dead. Croatian government officials wanted Owsley to help design a system for identifying the dead and determining their cause of death. For Owsley it meant helping families find out what had happened to their missing loved ones. And evidence of any atrocities would be forwarded to the international war crimes tribunal.

Owsley figured that the techniques and protocols he had relied on to identify scores of North American skeletons for the Smithsonian could be applied to identify the corpses in the Balkans. By studying the bones, Owsley could produce demographic data on age, sex, stature, and other physical characteristics on hundreds of war victims. The bone analysis would also shed light on cause of death. Owsley also advised the Croatians on how to inventory and code the deceased, and how to store and process the data on computers. The biggest obstacle Owsley foresaw was the lack of premortem records—documentation, photographs, or X rays from doctor or dentist visits that reveal injuries that would explain changes to a person's bones or teeth during his or her lifetime. While crucial for comparing and identifying the dead, most medical and dental records had been destroyed during the war. However, the Ministry of Health of the Republic of Croatia had been interviewing thousands of survivors and family members of the war dead in an attempt to obtain before-death characteristics of the deceased.

The idea of examining large numbers of war victims, properly identifying them, and creating an index was daunting. Owsley had no experience in such a large-scale project. But the trip to Croatia had helped him understand what would be required. Ready to return home, he had promised the Croatian government that he would return to their country and help train people to identify bodies.

Suddenly Owsley's attention was directed to the bright orange color on the television screen. An international news station was broadcasting a raging fire and billowing black smoke. An English-speaking commentator said something about a Waco, Texas, compound on fire.

Owsley stood up and raised the volume. The Branch Davidians, an American cult led by David Koresh, had been holed up in the compound since February 28, when agents from the Bureau of Alcohol, Tobacco, and Firearms (ATF) had raided the property in an attempt to arrest Koresh for weapons violations. The failed raid had left six people dead and led to a fifty-one-day siege. Then on April 19, the FBI used tanks to deliver tear gas into the compound in an attempt to end the standoff. The compound was soon engulfed in flames, with as many as eighty people believed inside. No one survived.

Owsley knew he would be needed back home.

Three days later
FBI Forensic Science Research and Training Center
Hairs and Fibers Laboratory
Quantico, Virginia

Seated at his desk, his wavy brown hair neatly trimmed and parted on the side, thirty-nine-year-old special agent Joseph DiZinno reviewed his latest research aimed at helping the FBI develop a means to extract and characterize DNA from hair. Trained as a dentist, DiZinno left private practice in 1986 to work as a criminal investigator doing dental comparisons for the FBI. Working alongside fingerprint experts, he developed evidence from crime scenes and helped positively identify both criminals and deceased crime victims through dental X rays, photographs, and the ridge counts in fingerprints and footprints. When the FBI decided to add DNA as a fourth means of developing evidence and identifying bodies, it asked DiZinno to act as a laboratory examiner overseeing mitochondrial DNA research.

Since beginning the research assignment in 1989, DiZinno had left the lab only twice to work on criminal investigations. In 1991 he worked on the kidnapping and murder investigation of the CIA's station chief in Beirut, Lebanon, William Buckley. And in 1992, DiZinno investigated the kidnapping and murder of Exxon executive Sidney Reso in Morristown, New Jersey.

He was about to leave for the third time.

The phone rang and DiZinno recognized the voice of Danny Greathouse, the chief of the FBI's Disaster Unit. Greathouse was in charge of the scene at Waco, and chiefly responsible for overseeing the removal and identification of all victims from inside the compound.

DiZinno asked how he could help.

Greathouse explained that there were a lot of bodies badly burned. They were going to have real difficulty in trying to identify them because of the fire and the ammunition it had set off. The compound had contained 1.8 million rounds of ammunition, 1.5 million rounds of which exploded in the fire. In addition to being burned beyond recognition, many victims' bodies were blown apart, leaving behind a human jigsaw puzzle of charred flesh and broken bones, many of which were severely fragmented by shrapnel. Used to handling disaster scenes resulting from plane crashes, train wrecks, and natural disasters, Greathouse had never seen such wide-scale human carnage in his fourteen-year career. With fingerprints an unlikely method of identification, he needed DiZinno's dental expertise.

Twenty-four hours later, DiZinno arrived at the Tarrant County Medical Examiner's Office in Fort Worth, Texas, where a team of medical examiners was receiving body bags full of human remains. Texas Rangers and FBI field agents had begun removing bodies from the Branch Davidian compound and transporting them the approximately one hundred miles to the medical examiner's office. Surviving family members were desperately calling to find out the fate of loved ones inside the compound, a virtually impossible task. The human remains in the body bags were so commingled that before identifications and autopsies could take place, bodies had to be put back together. And the medical examiner had only one anthropologist on staff. DiZinno thought he knew someone who could help.

When the FBI had been asked to help Ohio authorities solve a grisly murder in 1991, the authorities had brought in Owsley. A plastic bag with 286 bone and teeth fragments was sent to Owlsey at the Smithsonian. The bones were badly splintered, warped, bent, and mutilated. Nonetheless, Owsley identified the bones as the remains of Steve Hicks, an eighteen-year-old white male who had disappeared in the late 1970s. Owsley concluded that his bones had been cut and then broken by blunt force. After Owsley's analysis, Jeffrey Dahmer confessed to hitting Hicks on the back of the head with the rod of a barbell, then strangling him, dismembering his body with a Bowie knife and pulverizing his remains with a sledgehammer before scattering the bone fragments in the grounds around his parents' residence. Owsley's identification of Hicks led to the first murder conviction against Dahmer, for which he was sentenced to life in prison on May 1, 1992.

DiZinno picked up the phone and dialed Owsley at the Smithsonian.

April 27
Fort Worth, Texas
Tarrant County Medical Examiner's Office

"This is command central," DiZinno said, ushering Owsley through the Tarrant County Medical Examiner's office and into a conference room bustling with FBI agents wearing white shirts, ties, and dark slacks. Food provided by the Red Cross covered a table at one end of the room. Empty foam coffee cups filled a garbage can beside a cart with a VCR and television mounted to it. Videotapes of David Koresh holding children on his lap during the standoff played on the television screen. Banks of computers and phones lined the walls.

Working off a growing list of names of individuals believed to have been inside the compound before the fire started, agents telephoned immediate family members, attempting to acquire photographs, dental X rays, and medical X rays. The agents also asked whether the individuals had ever served in the military or worked for the government, a bank, or as a teacher—all professions that typically fingerprint personnel. In instances where possible sources of fingerprint cards were identified, FBI field agents in various parts of the United States fanned out to obtain them. The fingerprint cards were then shipped to the medical examiner's office, where FBI fingerprint examiners were prepared to verify a match. Since many of the victims were children and had not likely been fingerprinted, footprint cards were solicited from hospitals where Branch Davidian children had been born.

As DiZinno showed him the operation, Owsley immediately appreciated the burden he had inherited. With many of the bodies too dismembered to be identified by fingerprint analysis or to be compared to photographs and X rays, the FBI needed Owsley to put hands and feet with the right bodies, a process that would require him to sort out body bags full of twisted flesh and commingled bones. A systematic method of inventorying and matching remains had to precede any attempt to reconstruct people for identification purposes. Besides reconstructing the Waco victims, Owsley needed to generate a profile that included the age, race, sex, and height of each individual. That information, coupled with the premortem records being gathered by the FBI, would confirm the victims' identities.

Sizing up the task, Owsley decided to use the same research methods that he had learned under Bill Bass at Tennessee. As a graduate student, he had been to many burn scenes with Bass to recover charred remains. At that time, Bass, besides being the chair of Tennessee's anthropology department, was a deputy for the state of Tennessee's Medical Examiner's Office. Since Tennessee operates under a medical examiner system rather than a coroner system, every county has a physician—many of whom are in private practice—serving as medical examiner. The physicians were seldom available to visit crime scenes. Bass became an on-call medical examiner with statewide jurisdiction, employing the same techniques that he used to analyze historic skeletons recovered from archaeological digs.

Through trips to burn scenes with Bass, Owsley had learned how to identify charred human remains. With Bass, however, he had never been to a burn scene with so many bodies. He would need his team to join him.

The following day

Owsley's assistants Kari Sandness and Pam Stone followed the morgue technician into the walk-in freezer reserved for Waco victims. Inside, gurneys supporting black body bags lined the walls. Underneath the gurneys, body bags were stacked on top of one another. "When we receive the bags, we store them in here," the technician said. "The process is slow because we have to X ray every bag that comes in to make sure there are no live rounds in them." Once X-rayed, the bag would be turned over to Sandness for sorting. "Be real careful when you open the bags," the technician said. "If you find *anything* that looks suspicious or you're not sure what it is, tell somebody."

Leading them out of the freezer, the technician walked past the ambulance bays at the rear of the morgue to a garage with its door propped open a few inches. Inside, the walls were unfinished, lined with cinder blocks. Owsley stood between two rows of gurneys, each holding a black body bag. To Kari Sandness and Pam Stone it looked like a MASH triage unit. Both of them had seen their fair share of skeletons, but neither had been to a mass disaster site. It was grim business.

"Hi, gang," Owsley said, sweat beaded up on his brow. The garage had no windows, no vents, and no air-conditioning. The partially opened door and the large circular fans positioned strategically around the room to circulate air were hardly a match for the stifling ninety-degree heat that fueled the stomach-turning odor of rotting human flesh.

"This is gonna be pretty dirty work," said Owsley, his white lab coat smeared with mud and blood. He looked like a butcher. "It's not gonna be pleasant."

Listening, Sandness pulled on a white lab coat and latex gloves. Stone set up her black IBM laptop on a small metal surgery cart and plugged it in. "A lot of our time is going to be spent dealing with women and children who came out of the aboveground munitions bunker," Owsley said. "The human remains are so jumbled and mixed and incomplete that in a given body bag you might have four bodies." He let it sink in before he went on to explain.

Identifying a skeleton is relatively easy. But in this instance, the bones of each skeleton were separated and intermingled with others. They had to match bones together by identifying matching fracture lines and by relying on skin color, hair patterns, and socks, shoes, and other articles of clothing. They faced a giant task of matching.

"Basically, we're building people," Owsley said.

Few people could withstand even a snapshot of the horrific scene before Owsley's eyes, much less stare at it for days on end. The visual imprint that such carnage can leave on a person's memory would produce nightmares for many. But Owsley's focus was the daunting task of rebuilding the skeletons of so many blown-apart and burned people. His analytical faculties immediately became razor sharp, his senses and emotions all directed toward accomplishing his mission.

With Stone at the computer, Sandness slowly unzipped the first bag and gingerly pulled back the white sheet that lined the interior. Peering over Sandness's shoulder, Stone jolted her head back. A charred chest moved on its own. Maggots holed up in the chest's cavity created the illusion that it was "breathing," as it normally would when housing a properly functioning lung. "This is an active bag," Owsley said, looking on.

Sandness had previously handled skeletal remains, including some skeletons that still contained soft tissue. But she had never worked on so many children—children who were alive only days before. Delicately, she reached into the bag, then hefted out a mud-caked torso. One by one, Sandness removed human body parts, spreading them out on the gurney. The maggots started jumping and landing on her and Stone.

Suddenly Sandness pulled out a toddler-size red Keds tennis shoe. The toddler's foot was still inside it. "Now look for another child's foot with a red Keds sneaker on it," Doug instructed, before noticing that Stone's eyes were fixed on the shoe, her hands frozen at the laptop computer keyboard.

"This is hard-core forensics," Owsley said softly. "Lots of soft tissue. Lots of children. You need to numb these things out. You're doing a job. And you're doing the job to the best of your ability." Whether dealing with ancient skeletons or crime scenes, Owsley was primed to provide the facts to the best of his ability. He coached his assistants to do the same. "We don't have a vested interest," he told them. "We just have to learn from the remains and let them tell the story. We just need to be as factual as possible."

Sandness set the toddler's foot down on the table. A couple hours later, in a different body bag, she found the matching sneaker. It too had a foot inside it, making a match.

As the day wore on, a stray dog approached the garage door and stuck its nose under the opening. Stone and Sandness turned just in time to see the dog sniff and jerk its head back before scurrying off. "Even a stray dog doesn't want to be here," Stone said.

After days of sorting and matching the commingled bodies, Owsley had reconstructed the bodies of nearly two dozen women and children. With the aid of videotapes of the children taken shortly before the fire erupted,

Stone and Sandness were able to match more than one hundred tattered clothing articles—from shoes to diapers to hair barrettes—with the reconstructed children. Limbs missing and vital organs charred by fire and mutilated by shrapnel, the children and their clothing articles lined gurneys, ready to be inventoried by Owsley in preparation for autopsies. An identification number had been assigned to each victim.

With Stone at his side on the laptop, Owsley lowered his white medical mask over his nose and began with MC47C, the smallest victim recovered from the Waco compound. Arranged on a white linen sheet, MC47C stood out as the only victim without any recovered clothing items or remaining flesh. With a metal ruler, Owsley measured one of the femurs: barely over one inch in length.

"Based on the measurement of the long bone diaphyses, this is a fetus in about the fourth month," Owsley said, placing the tiny femur back on the gurney. "The fetal remains include the frontal, right humerus, radius, and ulna, left and right innominates, left and right femora, and left tibia."

The fetus had been recovered with a twenty-four-year-old female (MC47), whose long brown hair was found twisted around the fingers of another child (MC47B) recovered next to her. "Measurements of the radius and ulna suggest that the age of the child is approximately three and a half years."

Moving to the next gurney, Owsley stood in front of a body identified as MC73. "Age one point five," he said. "Body is in an advanced stage of decomposition. A foot found in a stocking was separate from the rest of the body. The infant is wearing a snap cotton undershirt and diapers saturated in urine, now crystallized. The waistband of the diaper is decorated with Disney characters. The child's pants are ankle-length, thermal underwear, with the brand name Young Stars."

"Doug, stop," Stone said, struggling to type fast enough to keep up.

"Tell me when you're ready," he said.

"OK."

"An extra pair of pull-up diapers was associated with this child," he continued. "The child is female."

The senseless killing of children angered Owsley. The only way he knew how to keep his emotions in check was to keep working. Shuffling his feet a few inches, he moved to victim MC74 without pausing. "This is a seven- to eight-and-a-half-year-old child," Owsley began. "The child is wearing dark cotton long underwear pants." With his index finger, Owsley cleaned off the clothing label. "The brand name is St. Michael," he said. "Maroon socks with rocket ships, gold stars, and the words 'lift off' embroidered on them. Black and purple tennis shoes with a basketball on them." He flipped the shoes over. "And the name Patrick Ewing is inscribed on the sole."

The next victim was a child less than two years old. "The diaper is filled with fecal material," Owsley observed.

Stone stopped typing. "Every diaper we've seen is dirty," she said. "Which means every kid was scared, scared shitless, quite literally."

"It's a natural human response when you're scared," Owsley said, practically whispering. "And there's no doubt these kids were in awful fear."

Owsley paused, sensing Stone needed a breather. In order to match clothing items with the bodies, Stone had spent days watching videos of animated children talking and playing—alive. She knew their voices, their names, and what they were wearing moments before they were hoarded into a munitions bunker loaded with grenades, .50-caliber machine-gun bullets, and heavy explosives. The compound had other underground facilities that were far removed from the munitions bunker and did not experience any fire. Yet the children were not brought there.

"Pam, how are you doing?" Doug asked, thinking momentarily about his own daughters back home as he watched Stone struggle. "Do you need anything?"

"No," she said. "It's just that you realize these people were making choices for these kids that didn't have a choice. They were making choices for little people."

Patiently, Owsley waited for her to put her fingers back on the keyboard.

"I'm ready," she said.

"You sure?"

"Yeah."

"The child had on white stockings," he continued, slowly, "folded down at the top, and size-five tennis shoes."

8

REMAINS OF THE DAY

Anxiety as to which body bag contained David Koresh increased as the number of unopened bags decreased. With only a few remaining in the medical examiner's walk-in freezers, rumors swirled that Koresh had escaped the fire and had been seen fleeing from the compound. Dentists working with the medical examiner ended the rumor, though, when they found a charred body, its legs and arms extensively burned in the fire. The body's skull was charred black and shattered into pieces, but the teeth were intact, and dentists were able to match them with dental X rays of Vernon Howell—aka David Koresh.

Medical examiner Dr. Nizam Peerwani asked Owsley to assist him in the autopsy to determine cause of death. It was apparent to everyone that Koresh had died as a result of fire, but Peerwani and Owsley weren't taking anything for granted. They knew that a shot fired by ATF agents had struck Koresh's hip region on February 28. Owsley had studied videotapes of Koresh holding small children on his lap inside the compound and vividly remembered Koresh wincing when one of the children accidentally sat against his hip. Sure enough, Peerwani and Owsley detected a fracture area with a marble-size circular hole in Koresh's hip. But it was quickly apparent that the bullet, although removing a large piece of bone, had missed Koresh's vital organs and did not cause his death.

The autopsy also revealed extensive charring on Koresh's legs, his left arm, and his genitalia. His head showed evidence of charring too, but the skull's extensive fracturing is what grabbed Owsley's attention.

While Peerwani completed the autopsy, Owsley took Koresh's skull fragments to a separate table for examination. Warped from burning, the size and shape of the fragments were distorted. Turning the pieces over in his hand, Owsley was intrigued that the edges were sharp and clear. This indicated some kind of trauma. Fire can cause skulls to fracture, but this did not appear to be a result of the burning.

His curiosity aroused, he spread out all twenty-eight pieces, which ranged in size from a quarter to a computer disk, in order to reconstruct the skull. Fitting and matching pieces together, he discovered a small round hole in the center of Koresh's forehead. A bullet hole, he immediately thought. When a bullet punches through a skull, it usually leaves a smaller opening on the exterior. Continuing to reconstruct the cranium with the aid of his two assistants, Owsley discovered a larger hole in the back of the head. It was the exit wound, he thought. Exits are usually larger openings, and there's external beveling. The back of Koresh's skull appeared as if something had been punched through the bone from the inside of the skull out. The hole at the back of the head also showed signs of charring and had fracture lines radiating away from it, both indications of a bullet exiting.

The wound in the forehead showed no signs of charring around it.

The evidence was clear. Koresh had been shot in the forehead.

Owsley stopped. Was it possible that he was dead *before* the fire? he wondered.

Owsley quickly reached for his calipers. As he began measuring the size of the wounds and the length of the fracture lines, he knew this would raise questions. Who shot him? Was it self-inflicted? Was it something the FBI did?

Suddenly he noticed an unusual characteristic in the bone around the exit wound. He was shot before he burned, Owsley realized. He either shot himself or he had someone shoot him before the fire had reached him.

Stone still didn't see how Owsley could tell.

He showed her how to read a bone. The skull, he pointed out, is made up of three layers: the outer surface, the inner surface, and a layer in between called the diploe. "You see this?" he said, pointing to the exit wound. All three layers of bone were visible. "The bullet that killed him traveled through the skull, exposing the inner surface and diploe layers of the skull to the exterior." He put the tip of his finger on the two internal layers of bone. "This is normally protected, internal bone, meaning it's inside the skull, not exposed," Owsley said. "The bullet has opened these inner bone layers to the outside of the skull."

The internal bone layers were blackened. The burning of the beveling told Owsley that the bullet came before the fire. These areas wouldn't be exposed to burning by a fire of this intensity unless a bullet first exposed them. The rest of the skull was only charred on the outer surface. No inner bone was charred—except where the bullet hole existed.

After carefully studying the trajectory of the bullet's path through the skull, Owsley finally speculated that Koresh did not shoot himself. It was more likely that he had one of his lieutenants, found dead with gunshot wounds near Koresh, pull the trigger. Owsley suspected that the man shot Koresh, then killed himself.

Owsley's findings became part of the final medical report that was turned over to the FBI.

Owsley's ability to convert the Waco death scene into a teaching situation impressed Joe DiZinno. To cope with human tragedy, many FBI agents, like police officers, resort to black humor at crime scenes. It's a sort of defense mechanism that divorces the humanity from the barbarism, enabling the law enforcement community to do its job. Owsley, however, never told morbid death jokes. He coped by trying to learn or extract as much information from a scene as possible. So DiZinno asked him to deliver a lecture to the FBI fingerprint team, thinking it might take the edge off of what had been a very difficult three weeks. Owsley agreed, and agents crowded into the morgue amphitheater to hear him.

An introvert by nature, Owsley could get instantly comfortable in front of large groups of strangers if he was talking about skeletons and the lessons to be learned from them.

DiZinno took his place at the rear of the amphitheater in the morgue, fingerprint experts flanking him. Owsley turned down the lights and flipped on the X-ray box light. One by one, he put up X rays of the remains of women and children he had just rebuilt with his team. Then he turned to face his audience.

"One adult was found in Bag B," he said. "She was lying facedown in the bag." He explained that he had opened the woman's badly decomposed hand to see if her fingers still contained fingerprints. "Her clenched right hand," he said, putting an X ray of it on the view box, "held the disassociated hand of a one-year-old child from Bag A."

"Whoa," one agent gasped.

Goose bumps shot up on DiZinno's arms as he and his colleagues stared solemnly at the X ray.

Owsley explained that at the instant the adult found in Bag B was killed, she was squeezing an infant's hand so tightly that she completely enveloped the smaller hand. But with both the woman's and infant's arms separated from the hands by the explosion, there was no clue to indicate the infant's hand was present inside the adult's hand. On the view box, tiny little fingers appeared under larger ones.

Only the gentle hum of an overhead-projector fan broke the silence in the amphitheater. FBI agents who thought they had seen it all shook their heads in disbelief.

When Owsley finished, DiZinno pulled him aside and informed him that one more task remained before he left Texas. All of the fire victims had already been removed, and the forensic scientists and most of the Texas Rangers had been released from the Waco compound. "But there are still four bodies that haven't been recovered," DiZinno said. "We've got information that leads us to believe they're buried in an underground bunker. The Texas Rangers have been down there digging and using cadaver dogs for days and have come up empty. Can you assist them?"

"Let's go look at the bunker," Owsley said.

9

THE PROBE

May 4
Subterranean firearms range
Northern edge of Branch Davidian compound

Driving from Fort Worth, DiZinno and Owsley's team were welcomed by the newly incongruous "Birthplace of Dr Pepper" billboard that announced they had arrived in Waco. The compound was still guarded by ATF agents armed with high-powered weapons. After verifying DiZinno's ID, the agents permitted DiZinno and Owsley's small team to enter the devastated compound. Some of the buildings were completely leveled. Others were reduced to ruins. Cleanup of the site was still ongoing.

Owsley trailed DiZinno toward the underground bunker, a concrete structure that, prior to the siege, had been used by the Davidians for target practice.

DiZinno looked at his watch. It was 4:00 P.M. Only a little over two hours of daylight remained. "I don't know if you're going to have any luck," DiZinno said.

Doug only nodded.

When they reached the thirty-foot-by-one-hundred-foot building, DiZinno rolled up the bottoms of his tan cargo pants, tucking the legs inside his rubber rain boots. During the siege, the Davidians began using the shooting range as a garbage dump. Much of the mud floor was obscured under a foot and a half of vegetable cans, fruit jars, milk jugs, and foam egg containers meshed with old clothes, glass, sheets of plywood, and hair from self-administered haircuts. Saturated with urine and stagnant rainwater, the garbage made it virtually impossible to breathe without the aid of military gas masks.

Owsley pulled a white biohazardous material–resistant Tyvex suit over his jeans and short-sleeve polo shirt. Looking up, Owsley noticed that the roof had been partially torn off after the siege ended. Sunlight streamed through the exposed rafters. A single lightbulb with a long white pull string was all that remained.

Two Rangers wearing gas masks and military fatigues and carrying shovels led Owsley and DiZinno inside. In the mud, Owsley observed paw prints and numerous holes dug by the Rangers. "Well, what have you been doing to find the bodies?" Owsley asked them.

"We've had a cadaver dog searching," one Ranger said, pointing to holes he had dug in spots where the German shepherd had signaled. "And we're getting this garbage turned up. The bodies are supposedly under this garbage. But we can't find anything."

Owsley leaned over and observed the paw prints more closely. "With this much debris and disturbance to the site, I can see how a dog might get confused," Owsley said, kicking back some loose trash. "Well, I might be able to help you. Have you got something to probe with?"

"Like what?"

"It needs to be something strong and sturdy."

"There's a piece of rebar in here," the Ranger said, pointing to a four-foot-long steel rod sticking out of the garbage heap.

"That'll work," Owsley said.

Skeptical, the Rangers leaned against their shovels as Owsley inserted one end of the rod in between some metal bars on a nearby tractor. Applying all his weight to the other end, he twisted the rod into the shape of a shepherd's crook. "I think we're ready," he said, holding the twisted end like a walking cane and staring straight up to the rafters.

Placing their gas masks back over their faces, the Rangers and DiZinno followed Owsley, maskless yet seemingly immune to the reeking urine and rain-soaked garbage as he walked to the back of the shooting range.

"OK, we need to have all this garbage pushed back to expose the ground," Owsley said. DiZinno grabbed a rake and began moving debris. Three Rangers followed his lead, peeling back the waterlogged trash into a pile. Within twenty minutes, the front section of the bunker floor had been exposed.

Working in a direct line, Owsley poked the soil. It was hard and compacted, indicating to him that the soil had not previously been disturbed. He stepped along the floor, plotting out a grid on the ground. Hardly hiding their boredom, the Rangers stood impatiently as Doug worked as if no one else were present.

After further clearing, Owsley started probing along the wall toward the far side of the room. Suddenly he stopped. Saying nothing and holding the probe with his left hand, he looked up at DiZinno. Grimacing, Owsley discreetly pointed straight down with his right index finger, flexing his wrist up and down.

Inching closer, DiZinno looked at his wristwatch: 4:30. Doug had been probing for about ten minutes. DiZinno thought Owsley had some special ability, as if maybe the rebar had become a divining rod.

The soil soft, Owsley pushed down harder with the pole, penetrating more than a foot below the surface. "This area's really spongy," he muttered.

Stunned, the Rangers moved in closer.

"This area's really spongy too," he said as he pushed the rod down again. Lifting the rod from the soil created a vacuumlike sucking sound in the mud.

Pushing the rod in the ground one more time, Owsley pulled it back. The hole he had created immediately filled up with a yellow, oily fluid, an indicator of excretion of human body fat and oils. He looked up at DiZinno. "Joe, this is it!"

Owsley backed up a couple feet, puzzling the Rangers as he made a series of holes that formed a rectangular outline to define the perimeter of the disturbed area. "This should be the grave shaft," he explained. "Let's dig here."

DiZinno and a Ranger started shoveling. Almost immediately they uncovered a large glove. Owsley picked up a shovel and joined them.

"There," Owsley said, pointing.

Everyone looked down. A red, mud-caked sleeping bag emerged from the soil. Zipped shut, it was difficult to tell whether the bag was blood soaked or made of red fabric.

"That's going to be a body right there," Owsley insisted, gently peeling the sleeping bag back. "See? You can see it."

"He found them!" one of the Rangers shouted.

Using his gloves to move the soil, Doug exposed a round object. "That's probably a face coming up right here," he said.

DiZinno and the Rangers looked on as Owsley exposed a badly decomposed human head. It faced upward. Decomposed soft tissue and a small amount of black hair were still present. Minutes later, the sleeping bag was removed, revealing an entire body in the supine position. The arms were extended downward; the legs were slightly bent. Argyle socks were on both feet. The individual, a male, was wearing blue jeans and a dark colored "Magbag" shirt over a T-shirt. A knee brace was noticeable under the jeans. Owsley noted the items with the body: a foam pillow, some 35-mm color slides in plastic slide cases, a bedsheet, and a wadded-up knit sweater.

Owsley paused. "It looks like there's another body under this one."

DiZinno resumed digging.

"It's a female," Owsley said. She too had been wrapped in a sleeping bag, her left hand by her side, her right hand draped across her chest. She wore glasses and a Star of David necklace. A watch was still strapped to her wrist. She was fully clothed, her feet clad in white high-top basketball shoes.

Over the next two and a half hours, the Rangers helped Owsley and DiZinno fully expose the grave shaft, which was four feet deep and contained all four missing persons. To Owsley it was as if they were stacked like cordwood.

Sweat running down his face, a Ranger rested on his shovel. "I don't mind diggin'," he said, looking at Owsley. "If they want me to dig holes, I'll dig all day long for 'em. But I'm gonna find that cadaver dog and shoot it."

Everyone laughed.

Outside the shooting range, DiZinno, drenched in sweat, removed his gas mask. "I would never have believed that you found these if I hadn't witnessed it myself," DiZinno said.

"Well, you learn to read the soil," Owsley said. "I was looking for differences in texture and soil density. I was looking for fresh disturbance."

DiZinno shook his head. "I'll always remember this day."

10

EVIL IS REAL

Three weeks later

Surrounded by stacks of papers on the dining-room table, Owsley attempted to finish writing his report on the trip to Waco.

"Dad."

"What, Kimmie?" he said to thirteen-year-old Kim standing at his side.

"Did you find any little kids down there?"

Owsley put down his pen.

"Yes."

"Were there any babies in there?"

"There were some babies too."

"What happened to them?"

Owsley paused, his memory capturing competing images of carefree children on videotape before the fire with those same children's burned, mangled bones after the fire. Over hundreds of hours, to accomplish his job, he had numbed his emotions to the evil and horror inflicted on the Branch Davidian children. He didn't want his daughter to know the horror either.

"Sweet pea," he said softly, "there were lots of little babies and kids that died. But I don't think we should talk about that."

"Can you tell me *anything* about the little kids?"

Owsley paused again. Since returning from Waco, he had kept his thoughts to himself. He had been reading the press reports that were critical of the federal government's conduct with relation to the raid. Owsley had a different view. He couldn't understand why Koresh, who had power over the children and responsibility for their safety, didn't release them. Instead, he had them put in a munitions bunker, where they were sure to die. "There were a lot of kids. And there were safe areas that were accessible by underground tunnels. But the children weren't put there. Instead, they were put in a place where they were sure to die."

"Why did they put the kids there? And why didn't the adults just take the kids out when the fire started?"

"There were some tunnels down there that they could have escaped through. But none of them got out." He hadn't answered her question.

Kim waited for him to say more.

"Now sweet pea, you need to go to bed."

Owsley blamed the children's fate on the adults—primarily Koresh— who had had control over them in the compound. Regardless of who was right, the government or the Branch Davidians, when push came to shove, the Davidian leaders should never have put the children in harm's way. Here, they were herded into the most dangerous place on earth—a munitions bunker. Normally when working a crime scene where a body has been dismembered, burned, or butchered, Owsley would take some satisfaction out of knowing his work would likely lead to the prosecution of the killer. Not in this case, however. Scores of innocent children had been killed, and no one would be held accountable. That bothered Owsley.

11

UNWRAPPING A MUMMY

Wednesday, May 8, 1996

On the Metro train commute from his suburban Virginia home to the Smithsonian, Owsley, always behind on his reading, retrieved from his bag a three-day-old edition of the *Washington Post*. He immediately zeroed in on the headline: "Nevada Mummy Caught in Debate Over Tribal Remains."

> For half a century, the Spirit Cave man lay in a wooden box with the lid screwed down tight, unforgotten but unexamined and uncontroversial. Then last month, anthropologists at the Nevada State Museum announced that new tests show the partially mummified skeleton is nearly 10,000 years old, and everything changed. The body appears to be one of the oldest found in North America . . . providing a rich view of life in the post–Ice Age world.
>
> Preliminary study strongly suggests the Spirit Cave man came from an ethnic group that long preceded the people whose descendants are the contemporary Indians of the Great Basin.

Owsley immediately wanted to see it. He figured he could solve the mystery over the mummy's ancestry and whether it was of American Indian origin. Since arriving at the Smithsonian, Owsley had developed the world's first comprehensive skeleton database. It contained profiles and descriptions of thousands of skeletons from scores of world populations.

Owsley got the idea for the database while teaching at LSU. One day Bill Bass called him from Tennessee and told him that the contents of the university's Over Collection—five hundred American Indian skeletons recovered from South Dakota by archaeologist William H. Over in the 1900s—were being repatriated to South Dakota tribes for reburial. Owsley wanted to document the collection before it went back. He asked Tennessee professor Richard Jantz to help him design a database that included information on place of origin, bone and dental inventories, demographics, skeletal and dental pathology, cranial and postcranial measurements, photographs and radiographs, and taphonomic observations. From the bones, Owsley wanted to be able to discern population demography, health conditions, migration patterns, patterns of gene flow, dietary habits, and mortuary practices.

Jantz was the natural choice to help design such a complex system. A master creator of databases, Jantz taught statistics and research methods at Tennessee. He also designed his own database for collecting skull measurements, a process that helped him identify skeleton populations based on cranial features. Owsley had thrived under Jantz as a student and had asked him to oversee his doctoral dissertation. An expert in dermatoglyphics—the analysis of fingerprints—Jantz suggested that Owsley analyze the fingerprint patterns of 204 Knoxville-area children with cleft lip or cleft pallet. Since fingerprints are a product of fetal growth and development, much could be learned by comparing the prints of afflicted children with those of their unafflicted relatives. After the study, Owsley produced a 182-page paper. The conclusion was that children with cleft lip and cleft palate did not suffer from markedly different rates of embryonic growth.

Impressed, Jantz asked Owsley to help him work on various research projects. Their crowning work together was the Over Collection project. Together, they inventoried all 500 skeletons, forming the foundation of their database. Since 1985, when they finished that work, they had added more than 4,500 other skeletons from a wide variety of human populations. As the data grew, they produced formulas that allowed them to identify any skeleton's identity by taking measurements and entering them into a complex formula of comparisons to those remains already compiled. If the Spirit Cave mummy's information could be entered into the system, a statistical profile would emerge, verifying whether or not it was of American Indian origin.

Owsley scanned the *Post* story for a contact person at the Nevada museum: curator Amy Dansie.

Before noon, he had Dansie on the phone. She explained that in 1940 two archaeologists working for the Nevada State Park Commission discovered a mummy—a five-foot, two-inch male in his forties—buried in a cave outside Fallon, a small town sixty miles east of Reno. It was unusually well preserved, the result of being sheltered in a cave in Nevada's arid climate. Based on artifacts found with the mummy, experts in the 1940s estimated it was roughly two thousand years old. Packed and stored at the Nevada State Museum, the remains received no attention until 1995, when advances in radiocarbon dating prompted the museum to have the mummy dated. The results showed he was approximately 10,650* years old. Dansie told Owsley that after the date came back, two Nevada Indian tribes—the Paiute and the Shoshone—claimed the mummy.

Listening to Dansie, Owsley's mind began churning. He and Jantz had just finished analyzing a series of skeletons housed at Brigham Young University and at the Utah Museum of Natural History. They had found that Utah Indian populations, from a skeletal standpoint, are quite different

---

* The Spirit Cave mummy date and all other skeleton dates in this book are based on calendar years, not radiocarbon years. Archaeologists and anthropologists measure time in radiocarbon years, arriving at ages through a complicated dating technique that measures the rate that the carbon-14 isotope decays. The Spirit Cave mummy's radiocarbon date is 9,430 years old. But the level of carbon-14 in the atmosphere can vary year to year. Hence radiocarbon years do not equate to calendar years.

Stafford Research Laboratories in Boulder, Colorado, performed the calibration from radiocarbon years to calendar years for the skeletons featured in this book. The calibration from radiocarbon to calendar years produces an age range. For example, Spirit Cave mummy's radiocarbon date of 9,400 years calibrates to between 10,750 calendar years and 10,550 calendar years. The dates featured in this book reflect the midpoint of the range; hence 10,650 years for Spirit Cave mummy.

from Plains Indians. Unlike Plains Indians, who tend to be quite big, the Utah American Indians are much smaller in physical stature.

Dansie doubted that the mummy found in Spirit Cave had any affiliation to either the Paiute or the Shoshone.

Owsley assured her that by coming out and taking some standard measurements, he could confirm whether the Spirit Cave mummy was a descendant of the Paiute or the Shoshone, or whether he was unrelated. So, it was decided; he would visit Dansie sometime in July.

Owsley could not wait to get to Nevada. Studying prehistoric skeletons contrasted sharply with the work he performed on fresh bodies from modern crime scenes and mass disasters. All of it was part of his job, but the historic remains were less like work than the modern ones.

Whenever a law enforcement agency called Owsley, the news was bad. He never got called when people died naturally. His cases were always ones in which humans had been so badly disfigured or decomposed that traditional forms of identification were inadequate. On the other hand, good news and a sense of excitement and intrigue accompanied the discovery of older remains. The bones of the distant past had no fresh sense of death, nor any mourning relatives. Owsley saw these skeletons as a chance to learn and enlighten the future.

Given a choice, Owsley had no preference, however. He viewed the modern cases as a challenge where science could trump evil. By identifying a maimed or dismembered body, he could deprive a killer of his design to obscure a victim's identity. Also, Owsley could lend solace to the survivors and evidence to assist law enforcement officials in the prosecution of perpetrators. These were the tangible payoffs to his forensic work.

Conversely, the study of historic skeletons was simply knowledge for knowledge's sake. Few people in the world appreciated skeletons as Owsley did. Since bones were like books to him, Owsley's encounters with ancient skeletons were akin to a contemporary playwright stepping into a time machine and spending a day with Shakespeare. Only Owsley had the luxury of getting into the time machine over and over again.

Nonetheless, he felt obligated not to confine his work to historic remains. That would be selfish. He knew he possessed an innate ability to perform a valuable service after tragic deaths. Besides his scientific skills, he had trained himself to numb his mind to the horrific scenes before his eyes. He could work on Branch Davidians and sleep at night. His dreams, which were often in elaborate color, never involved the skeletons he worked on. Yet he recognized he could never be a full-time medical examiner, confined to handling recent death cases all the time. He needed to see the older ones. To him, they had stories to tell, stories that could only be interpreted through their bones.

Carson City, Nevada
July 15, 1996

Tall, with bushy graying hair, sideburns, and glasses, fifty-three-year-old Richard Jantz stood over the wooden observation table with Owsley. They watched as Smithsonian photographer Chip Clark placed an index card with black bold letters that read SPIRIT CAVE. Clark set it on a white cloth sheet in front of a wickerlike matting made of intricately handwoven reeds. A small hole in the matting revealed a section of the Spirit Cave mummy's articulated spinal column. Clark rubbed his thick black beard and adjusted his glasses. "I'm ready," he said, camera strap around his neck. As the morning progressed, he would shoot the black-and-white film to be archived at the Smithsonian, where it was expected to endure for centuries. The color slides he took would be scanned into a digitized computer database, available for publications and classroom presentations. Together, the two photographs assured a permanent visual record of the oldest mummy in North America.

Curator Amy Dansie, her graying hair pulled back in a ponytail, gently drew back the reed matting, revealing a second layer of diamond-plaited matting, intricately hand-stitched and sturdy. The sight of diamonds from over ten thousand years ago gripped everyone. Owsley, Jantz, and their assistants looked on silently from the opposite side of the table.

Lying on his right side, the mummy had his right arm flexed at the elbow, his wrist resting under his chin. The left arm was extended in front of the pelvis. Some joints contained dried cartilage. Ligaments were attached to bones covered by desiccated skin. The spinal column wound toward a robust set of hips, above which rested portions of the large intestine and colon. Both legs were slightly flexed. A pair of leather moccasins covered the feet.

"How could something so old be so well preserved?" marveled Jantz.

"Wow," Owsley whispered, zeroing in on the mummy's skull, mostly covered in desiccated, leathery brown skin and scalp tissue, from which clung brownish red hair that hung down the sides of the head. Two patches of hair dotted the top of the skull. Visible through a small opening in the cheek, brown-stained teeth sat firmly in the jaw. Recessed and round, both eye orbits had eye tissue and muscle attached.

Owsley bent down and cocked his head, his eyes level with the mummy's face. "Well, look at that," he said, peering through the empty eye cavities into the skull. Strands of tissue resembling spiderwebs traversed the interior cranium.

Slowly, Owsley stood back up. "He has a short, narrow face," said Owsley, thinking privately that the mummy seemed very different from the characteristics he was used to seeing in American Indians. They typically have wide faces. "And the nose and chin jut forward," he continued, thinking that those features were also atypical.

Owsley began inventorying the bones. He stood over a circular halogen lamp and, through a magnifying lens, looked at the skull.

"He's got a skull fracture . . . a depression fracture right here on the side of his head," Owsley said, using a dental pick to point to a depression in the left temporal area behind the mummy's left eye. He figured the man had been hit with a rock or a blunt weapon, with sufficient force to leave a dent in the skull the size of a quarter. Owsley rotated the skull in his hands. Crooked fracture lines traveled away from the depression, one going up and the other going down. The bone around the depression was slightly darker in color, indicating that he had been in the process of healing when he died.

He felt the man's long strands of wavy brown hair with reddish tones. In this instance he felt confident red was not this person's natural hair color. After death, hair color can lose its pigment if exposed to natural light. Owsley figured the man's natural hair color was darker. Using his fingers, Owsley worked through the coarse hair, searching for nits and head lice. He found none.

Next Owsley looked at the spinal column. "He's got thirteen thoracics," he said. "That's an anomaly." The human anatomy typically has twelve vertebrae, not thirteen, a rarity passed on genetically. "And see, this is a fatigue fracture," he said, pointing to a fracture line in the spine. He explained that the fatigue was the result of chronic strain to the back, potentially caused by constant bending over to hunt or find food. Years of stress on the lower back ultimately fractured one of the vertebrae.

"There is some intestinal content," Doug said, pointing to a portion of the large intestine that remained intact and stretched to the colon. It held dried feces, a potential gold mine of information that would reveal the mummy's last meal.

When Owsley finished with the skull, Jantz set it on a cushioned, doughnut-shaped ring that was suspended six inches above the table by a metal rod. Once the skull was secure, Dansie reached her fingertips gingerly under the brittle skin. Peeling back small areas covering specific landmarks on the skull, she gently exposed the bone surface.

For the next two hours, Jantz took more than fifty skull measurements, capturing the facial forwardness, vault breadth, facial breadth, and facial height. It was clear that the skull had a long, narrow vault, very atypical of Native Americans. It was not a skull with a vault morphology that he would expect to see in modern people in the area. It was very uncommon for what one might find in the Great Basin or in any Indian in the past three thousand years.

Jantz handed the skull to Dansie. Owsley and Jantz gathered around her as she returned the skull to the mummy's other bones.

"This is very interesting," Owsley said, taking a final look at the Spirit Cave man before Dansie covered him back up in his diamond-studded matting. "This guy is so fascinating, so different. He doesn't look like anybody alive today."

He could hardly wait for Jantz to run Spirit Cave mummy's cranial measurements through his computer program that would compare him to other human populations.

12

SOMEBODY ELSE IS HERE

"Doug, it's Richard."

"Hello, Richard. What did you find?"

"This person is not only dissimilar to Native Americans, he wouldn't fit very well in *any* modern population."

Jantz felt that the Spirit Cave mummy most closely resembled the Ainu, a maritime people that anciently occupied coastal Asia. Survivors of the Jomon culture, which dates back at least ten thousand years, the Ainu have survived to the present day and still occupy Hokkaido Island, in the northern part of Japan.

Jantz explained that after the Ainu, the Spirit Cave mummy shared some similar features with the Moriori—a Polynesian group—and the African Zulu, as well as the Norse, a medieval Norwegian population. The population groups that least resembled Spirit Cave mummy were the Bushmen, the Sioux, and the Pawnee.

Somebody else is here, thought Owsley, listening to Jantz. The Spirit Cave mummy's cranial measurements were radically different from any of the thousands of Indian crania he and Jantz had studied.

"So he doesn't necessarily look like an Indian at all," Owsley said.

"Well, we knew he looked different. This just confirms it."

Owsley was intrigued by the mummy's connection to the Ainu. But it raised a question, and a loaded one at that: How did the Spirit Cave mummy get here? Owsley realized that it posed a challenge to the prevailing archaeological theory on how the Americas were populated.

Since the 1950s, most experts have believed that the first Americans arrived sometime around thirteen thousand years ago via a land bridge that linked Siberia and Alaska at the Bering Strait. The theory rested partly on archaeological discoveries at Clovis, New Mexico, in the 1930s. Stone tools and artifacts found at Clovis gave rise to what scientists dubbed "Clovis people," believed to be big-game hunters who crossed the land bridge and migrated south along an ice-free corridor that formed after the Ice Age. But no skeletons or human remains were found at Clovis. Yet supporters of the Bering Strait theory regard Clovis people as the forefathers of modern-day American Indians.

Yet scientists had been unable to confirm the presence of a single Clovis skeleton at any of the forty additional Clovis sites uncovered by archaeologists in North America since 1930. While the tools and artifacts matched those at Clovis, New Mexico, lending strength to the existence of the Clovis culture, a profile of the Clovis people remained elusive.

Owsley and Jantz agreed that they needed to make a comprehensive inquiry into how many other skeletons from the Spirit Cave time period existed in America.

If the Spirit Cave mummy looked nothing like North America's Indians, Owsley thought, it raised the question of whether another migration to the Americas took place.

Jantz already had doubts about the single-migration theory. He had argued that the isolation that historically has been postulated for America—they came over the land bridge, the land bridge disappeared, and thereafter they were in exquisite isolation—cannot be the case. There is too much evidence of interaction.

Owsley did not believe that a single migration was responsible for populating all of North and South America either. He suspected that there were multiple migrations to the Americas. And, despite never saying it publicly, he believed that there was probably at least one ancient population that migrated to the Americas by boat.

The Spirit Cave mummy's apparent connection to the Ainu got him thinking more about a boat migration. The Ainu were, after all, a maritime people.

Owsley and Jantz agreed that they needed to study more of these early mummies and skeletons. Trying to build the case that the early people in America really were different peoples would require them to explain why. That would involve going beyond individual specimens. The Spirit Cave mummy was an important piece of evidence. But no scientist could ever build a case for his theory on one specimen.

13

HUMANS REMAIN

———————————

July 28, 1996
Kennewick, Washington

"I think I've almost got the route we're going to take all figured out," Floyd Johnson announced, appearing in the bedroom doorway, his putter in hand. He had just come in from the front yard, where he had been chipping golf balls.

"Honey, did you remember to get the oil changed on the car?" Suzanne asked, knowing what his response would be. Although their annual three-week vacation was still four days away, Suzanne packed early, eager to get to Southern California.

"I'm going to do that tomorrow," he said, grinning sheepishly. After placing his putter back in his golf bag, Floyd reached for the atlas. "Suzanne, did you pack me enough underwear?"

She clenched all twelve pairs in her hand and held them up over her head. "While you were practicing for the PGA out in the front yard, I packed them for you."

"Well, you are doing such a fine job at packing, maybe I'll just leave the rest to you and go take a few more shots," he said.

"I have a better idea," she said.

"What's that?"

"Why don't *you* finish packing and I'll go sit in the car and honk."

Suddenly the phone rang. They both looked at the clock. It was nearly 5:30 on a Sunday afternoon.

"I'll get it," Floyd said.

Suzanne did not have a good feeling about this.

Johnson recognized Kennewick Police sergeant Craig Littrell's voice immediately.

"We need you to respond to Columbia Park," Littrell said.

"What's up?"

"A human skull has been recovered," Littrell said. He explained that two boys attending the Tri-City Water Follies boat race found it. They were wading in the Columbia River just off the shoreline when they discovered the skull, which was submerged in the river. There were additional bones found on the shoreline that appeared to be human.

"I'll be there in a few minutes."

Since retiring as a homicide detective years earlier, Johnson, fifty-seven years old, had served as the Benton County coroner. Every time a death occurred within the county, unless the deceased had been in the care of a physician thirty-six hours prior to death, Johnson responded to the scene, examined the body, and determined whether the cause of death warranted a criminal investigation.

Johnson hung up and returned to the bedroom, silently placing the atlas back on his nightstand. Recognizing the solemn expression on his face, Suzanne knew someone had died.

"Sweetie pie, I gotta go," Floyd said. "They got one down in the park."

"How long do you think it will take?" Suzanne asked as Floyd changed into a pair of long pants.

"I don't know," he said, fastening his belt as he walked out to the driveway toward his brand-new green four-wheel-drive Jeep Cherokee. The white lettering on the side said, BENTON COUNTY FOR OFFICIAL USE ONLY. Inside, a dozen white body bags were neatly packaged in plastic on the backseat, each one bearing a warning notice: "Use universal blood/body fluids precautions with all patients."

The Jeep held everything he needed: his black leather medical bag, a box of disposable latex gloves, flashlights, a 35-mm Canon camera, and a Polaroid camera. He also kept a Smith & Wesson .380 automatic pistol under the seat.

Johnson turned the ignition. His black Kenwood police hand radio came on with the engine. Backing out of the driveway, Johnson glanced at the tray under the glove compartment, double-checking for the blue jar of Vicks VapoRub. When going to a death scene that had a decomposed body, a dab of VapoRub under the nose was his best defense against the odor.

Johnson drove five miles to the Columbia River. Spectators from the boat race were congregated around an area cordoned off by yellow police tape. Inconspicuously dressed in blue pants, a short-sleeve green pullover shirt, and black shoes, Johnson slipped under the tape and found Sergeant Littrell.

"Dr. Death is here," another officer joked as Littrell handed Johnson a white plastic bucket.

Johnson peered inside. At the bottom, a skull rested in a puddle of muddy water. A dark yellow residue covered its surface. Two ominously empty eye cavities in the cranium stared up at him.

Johnson lifted the skull, mud and water oozing out. "This is pretty heavy," he said, rotating it in his hands. He observed a fracture line just below the cheekbones, about where the bridge of the nose was. Johnson reached back into the bucket and pulled out a plastic bag. It held two pieces of the jawbone. The top and bottom teeth were still intact.

Littrell told him that a number of other bones had not yet been removed from the river.

Silently, Johnson stared at the yellowish brown streaks on the skull, convinced that the skull had to be over a hundred years old. He had seen his share of murder victims whose remains had been reduced to skeleton. But these bones looked different from anything he'd seen before. He suspected the skull was from a deceased Indian. Numerous tribes lived in the area, and ancient burial grounds were plentiful in the river basin region. It wouldn't be the first time that river erosion had exposed Indian remains.

Gingerly, he placed the skull back in the bucket. "This looks *old* to me," Johnson told Littrell. "But I want to have an expert look at it."

The bearded 47-year-old Dr. James Chatters met Johnson at the doorstep to his house. Chatters, a paleontologist in Richland, worked as a consultant to the county coroner's office and to numerous Native American tribes. Saying nothing, Chatters knelt down under the large white birch tree in his front yard and performed a quick forensic study of the skull. He checked the length of the skull, the shapes of the seams where joints of the skull came together, the prominence of the chin, the structure of the jawbone, and the features of the other, smaller facial bones. Then he gently placed the skull on his green lawn and lifted the jawbone from the bucket. After looking at it, he put it back down and lifted up the skull again.

"This is *very* unusual," he whispered. To him it looked like an old European skull.

"I thought it was probably an old Native American's remains," Johnson offered.

"No, this is not Native American," Chatters said. To him there was nothing Native American about it. There were unusual characteristics in the skull, narrow, sloping features. And the prominent nose, distinct from the more rounded, Mongolian-type features found on early Native American remains.

Chatters decided it would be best for him to see the discovery site.

It was dusk when they reached the river. Chatters went directly to the riverbank with a police officer to see if any bones were visible onshore. None were. Quickly walking the beach, Chatters found various artifacts: white ceramic, a horseshoe, a piece of wood with old square nails stuck in it, and a cow bone.

The dive team ready, Chatters and Johnson boarded the Benton County Sheriff's rescue boat and trolled fifty yards offshore, where they began circling within a two-hundred-yard radius. Using flashlights, the team spotted more bones in the shallow water, protruding from the mud on the river floor. Chatters hastily removed his leather sandals and climbed overboard. Feeling a hard surface against his toes, he bent over, reached into the muck, and lifted a piece of pelvis bone. He handed it up to Johnson. Next he located another piece of the lower jawbone, more pieces of the pelvis, a femur, bones of the lower leg, and then pieces of the upper arms.

Convinced the media and curious spectators would overrun the site once news of the discovery leaked out, Chatters groped along the riverbed until it got too dark to see. Before climbing back into the boat, he had retrieved dozens of bones to go with the skull. Noting their position in the river, Chatters figured the skeleton had recently washed out of the riverbank because of erosion. Few of the bones contained algae, and even fewer of them showed signs of surface erosion. The bone matter was in extremely good condition, something rarely seen in recent materials.

Chatters concluded from the physical characteristics of the skull and the fact that there were a lot of historic artifacts strewn on the beach from around the turn of the century that they probably had a skeleton from the early settlement era, the early 1800s, when Lewis and Clark camped nearby. It was presumably somebody who had been buried in a family plot outside a homestead. Often people were buried not far from the house.

After helping Chatters load the bones and other objects into the back of his Jeep, Floyd pulled out of the park, passing a historical marker overlooking the Columbia River. He had passed it countless times before, never paying particular attention to the wood sign, it being intended to aid tourists unfamiliar with the spot's significance. The words LEWIS AND CLARK EXPEDITION were engraved in yellow across the top.

"The large island seen from here marks the furthest point upstream in the Columbia River reached by the Lewis and Clark expedition, on October 17, 1805," the sign read:

During the encampment of the party at the mouth of the Snake River, Captain William Clark with two men ascended the Columbia in a small canoe. They found the Wanapum Indians, who lived in mat houses along the shores, engaged in drying large quantities of salmon.

"The multitudes of this fish are almost inconceivable," they reported. "The water is so clear that they can readily be seen at a depth of 15 or 20 feet." At one of the houses visited, a boiled salmon was served to each explorer. Joining them in 18 canoes, the Indians pointed out the mouth of the Yakima River which they called the Tapteal. As they were making the return trip downstream, Captain Clark shot a sage grouse that measured 42 inches between wing tips.

Just after 9:30, Johnson arrived back home. The suitcases on his bed had been moved to the floor. His golf bag was in the corner. "You were gone a long time," Suzanne said.

Seated on a stool in his laboratory, Chatters carefully cleaned and sorted the bones that he had laid out the previous night to dry. Using a soft brush, he removed mud and residue from the exterior surface. Suddenly, Chatters squinted, raising the pelvic bone closer to his eyes. A gray object appeared embedded in it.

Realizing it was a potential clue to the skeleton's age, Chatters immediately took the pelvic bone to Kennewick General Hospital and had it X-rayed. But the X-ray equipment failed to detect the projectile. It indicated to Chatters that the object was not metallic.

His adrenaline rushing, Chatters had the pelvic bone CAT-scanned. The more precise computerized image homed in on the object, revealing a spear point with a rounded base and serrated edges. Chatters thought the original wound appeared to have healed over the spear point, then reerupted.

He could scarcely believe his eyes. A stone spear point lodged in a Caucasoid skeleton conflicted with traditional views of American history. Outside some of the battlefields in the Midwest, researchers hadn't found stone points stuck in European settlers. Was this an early Kennewick pioneer who got in trouble? Or perhaps he was an explorer who was not on record? That would be a historical discovery.

Seeking a second opinion, Chatters took the skeleton to Catherine J. MacMillan, a physical anthropologist at the Bone-Apart Agency at

Central Washington University in Ellensburg. While she closely examined the skull and its other bones, Chatters waited outside her office. Finally she emerged.

"Caucasoid male, forty-five to fifty-five years old at death," she said matter-of-factly.

Her conclusion supported his. Chatters then handed her the pelvic bone, which he had withheld from her during her initial observation. He pointed to the stone object embedded in it. "Does this change what you think?" he asked.

Stunned, MacMillan stared at the stone projectile. "Well, how can that be?" she said, asking to see the skull again.

Silently, she reexamined it. No wide cheekbones, she observed. No shovel-shaped incisor teeth.

She picked up the pelvic bone again. "It's interesting that it's in there," she said, staring at the stone projectile. "But my opinion remains the same—Caucasian male."

Chatters hurriedly returned to his office, eager to call Floyd Johnson, who was due to leave on his three-week vacation in less than twenty-four hours. Chatters needed Johnson's authorization for what he wanted to do next: send a bone sample to the University of California, Riverside, for radiocarbon dating.

Reached at his office, Johnson weighed Chatters's request. He knew the remains were found proximate to an area traditionally known as a Native American burial ground. But two scientists had independently determined the skeleton could not be Native American. And the stone spear point suggested the skeleton predated the Lewis and Clark expedition. They were potentially sitting on a very important historical find.

Johnson approved Chatters's request.

August 26, 1996

It was 8:30 A.M. when flimsy white fax paper started scrolling from Floyd Johnson's machine. The letterhead alerted him that the transmission was coming from the AMS Research Facility at the University of

California, Riverside. Johnson had been back a week from vacation, and he had thought often about the case while away.

"Dear Mr. Johnson," the letter began. "The results from the Columbia Park remains, designated as radiocarbon sample APS-CPS-01 are as follows." Intrigued, Johnson scanned down the page, looking past a complicated chart with numbers and formulas. His eyes stopped at the heading "Calibrated Age." Beneath it he noticed the date: "7265 B.C.–7535 B.C."

Confused, Johnson searched the page for further explanation. His eyes zeroed in on a number: "9,800 years." He looked back up at the date: "7265 B.C.–7535 B.C." Johnson kept reading. "The calibrated value is expressed with a 95% confidence level in years B.C.," the letter said.

Motionless, Johnson looked up from the fax. The skeleton in his custody was approximately 9,800 years old. Never mind Lewis and Clark or Christopher Columbus; the man in Jim Chatters's basement laboratory seemed to predate every documented account of people migrating to the Americas. The date raised more questions than answers. Where did he come from? What was he doing in North America back around the time of the Ice Age? And how did he get here?

Unsure of the implications, Johnson wanted to call a press conference the next day and announce to the world that Kennewick, Washington, was home to North America's oldest human. The skeleton would be aptly named: the Kennewick Man.

14

AIRFARE FOR A SKELETON

"Treat the remains as Native American until proven otherwise." That's what the Umatilla Indian tribe had told the Army Corps of Engineers back on July 29, the day after Kennewick Man surfaced in the Columbia River. In 1855, the Umatilla ceded lands to the United States in a treaty that formed the Confederated Tribes of the Umatilla Reservation. The treaty established a reservation in Oregon, approximately seventy-five miles south of where Kennewick Man surfaced. Prior to the treaty, the Umatilla were among a number of tribes that inhabited or used the Columbia River area around Kennewick.

Before the Kennewick Man discovery, both the corps and the local coroner's office had found Indian remains along the Columbia River and turned them over to tribes for reburial. The Umatilla's religion requires that exposed remains be treated with utmost respect and immediately reinterred. To the Umatilla, any old remains were Indian remains. The tribe claimed that its oral and religious traditions held that the Umatilla did not have contact with non–Native Americans until 1805, when Lewis and Clark arrived.

The Umatilla's directive to treat Kennewick Man as if he were Native American was issued to the Walla Walla, Washington, office of the Army Corps of Engineers, which had oversight of the federal land where the discovery occurred. The same day that the corps heard from the Umatilla, corps archaeologist Ray Tracy talked to Jim Chatters, who said that the Kennewick Man looked nothing like the Indian remains that typically surface in the Columbia River region. Tracy issued to Chatters an archaeological resources permit authorizing him to return to the site in search of more bones and artifacts that might aid in establishing Kennewick Man's identity.

The Umatilla were not told that an archaeological permit had been issued to Chatters. Assuming no research or study would be performed on Kennewick Man, the Umatilla fully expected to take possession of him once the coroner's office formally announced that the discovery site was not a contemporary crime scene. The corps did not tell the Umatilla that Kennewick Man had been in Chatters's basement laboratory for five weeks; that it had been analyzed by a second anthropologist, who had determined that the remains were not Native American; or that a bone sample had been sent to California for radiocarbon dating.

Euphoric over Kennewick Man's age, Jim Chatters faced a quandary. Convinced that the skeleton in his basement deserved long-term study and analysis by the nation's top scientists, he sensed there wouldn't be time for either. Familiar with the Umatilla's oral tradition, Chatters knew that the tribe would demand Kennewick Man the minute Floyd Johnson announced the skeleton's age to the press. While he was convinced that Kennewick Man could not be Native American and should not go to the tribe, Chatters felt that he and Johnson might not be able to prevent it. As the titleholders to the federal land where Kennewick Man had surfaced, the federal government had the final say in the skeleton's destination.

Seeking advice, Chatters telephoned Professor Gentry Steele, an anthropologist at Texas A&M who had studied and published on prehistoric skeletons.

Listening to Chatters describe Kennewick Man's features, Steele agreed that it sounded like nothing he had encountered. He wanted to inspect it himself. But he couldn't get up to Washington in the foreseeable future. And the political situation that Chatters sensed arising gave Steele the impression that an expert in ancient remains should look at it right away.

"Whom do you suggest?" Chatters asked.

"I'd try Doug Owsley," said Steele

Chatters had heard of Owsley, but was immediately turned off at the prospect of involving the Smithsonian, which housed more Indian remains than any other museum in the country, a fact that angered many tribes. It would only inflame the situation with the Umatilla.

Steele told Chatters that Owsley had studied thousands of skeletons, including collections from many western tribes. And he explained that Owsley had just returned from the Nevada State Museum, where he had helped curator Amy Dansie analyze a 10,650-year-old mummy—one nearly the same age as Kennewick Man. "Doug has a whole team that comes in to do it," Steele said.

Chatters hung up with Steele and decided to sleep on the question of whether to call Owsley. Besides trying to avoid the Smithsonian's negative image among Indian tribes, Chatters also did not want some hotshot Smithsonian scientist taking away his opportunity to study Kennewick Man.

The next morning, before the press conference, Chatters called the Nevada State Museum. He reached Amy Dansie. After explaining her situation and the service Owsley and his team had performed at her museum, Dansie gave Chatters the same advice Steele had: Call Doug Owsley.

Instead, Chatters went to a hastily organized press conference at city hall. It drew very little press: one local newspaper reporter and two local television reporters. Following a brief explanation of the date and its relevance, Chatters and Floyd Johnson sat next to each other at a press table and fielded questions about Kennewick Man's origins, unusually well preserved bones, cause of death, and population affiliation. "We don't know," Chatters responded more than once.

Right after the press conference the Umatilla called the Army Corps of Engineers. The tribe's religious leader, Armand Minthorn, was indignant that one of Kennewick Man's bone fragments had been sent to California for radiocarbon dating without his permission. Minthorn trusted the radiocarbon date's accuracy. In his eyes, however, the test was unnecessary and unwarranted. It only confirmed what Minthorn and the tribe had said five weeks earlier when the discovery was made: Kennewick Man was ancient and any ancient skeleton was a tribal ancestor. "Our religion tells us so," Minthorn said. "Our oral history tells us so. All of those tell us that we were created here. We did not cross any land bridge like the scientists tell us. Our religion tells us we were created here. Period."

Minthorn and the Umatilla put the corps on notice that other tribes from the Columbia River region were angry too. The Umatilla had mobilized the Yakima, the Wanapum Band of Yakima, and the Colvilles from Washington, as well as the Nez Percé from Idaho. Prior to 1855, they had been united. That year, the United States government divided and renamed them, assigning them to separate geographical land bases in Washington, Oregon, and Idaho. Today, the five tribes form a confederation. Together, they applied immediate pressure on the Army Corps of Engineers to repatriate Kennewick Man for reburial.

Thursday, August 29

Asked by the corps to attend a meeting with tribal leaders, Chatters arrived flanked by corps archaeologists Ray Tracy and John Leier. They approached the spot where Kennewick Man had surfaced five weeks earlier. Armand Minthorn and four other leaders stood waiting for them. Chatters recognized a couple of them, with whom he had good relations from previous consulting work he had done for their tribes. Lieutenant Colonel Donald Curtis, the commanding officer of the corps office responsible for overseeing the disposition of Kennewick Man, had sent his executive assistant, Lee Turner, to represent the corps' interests. Tall and adroit, Turner apologized to the tribal leaders for Lieutenant Colonel Curtis's inability to appear in person.

While Chatters and the corps officials shook hands with Minthorn and the other tribal leaders, two more Umatilla Indians arrived. Chatters recognized one of them immediately—Jeff Van Pelt, the tribe's cultural resources manager.

Van Pelt immediately criticized Chatters for having a bone sample from Kennewick Man age-tested. He accused Chatters of knowing the remains were Native American.

Turner quickly took control of the meeting by assuring the tribal leaders that he had full authority to speak for the colonel. He then asked to hear from the tribes. Minthorn stepped forward to speak, saying that the remains should not have been disturbed and ought to be immediately reburied.

Turner assured Minthorn that his wishes reflected the colonel's top priority.

Additionally, Minthorn wanted no further study or publicity around Kennewick Man, to which Turner also offered his assurances.

Chatters bristled at Turner's willingness to stifle study or publicity and suggested that the U.S. Constitution prohibited the kind of cover-up they were agreeing to. But he suspected that by himself he couldn't stop them, and suddenly he had second thoughts about not calling Doug Owsley. As soon as he returned from the site to his office, Chatters telephoned the Smithsonian.

"This is Doug."

Chatters introduced himself and gave Owsley the lowdown on Kennewick Man.

"That's very interesting," Doug said. "The morphology you're describing sounds like a mummified skeleton I just worked on in Nevada at the Nevada State Museum, called the Spirit Cave mummy."

"Yes, I'm a little familiar with that. I spoke with Amy Dansie at the museum yesterday. She told me about the Spirit Cave mummy and that you had been out there to look at it."

"What makes Spirit Cave so different is that with Indians you have long faces and short, broad crania," Owsley said. "But the Spirit Cave mummy has a short face and relatively long cranium, a totally different constellation."

"Yeah, that's the way this guy looks," Chatters said.

Owsley wanted to see it.

"I have a grant to conduct archaic studies in the West. I could probably get my team out there in a few weeks," Owsley said.

"Ah, I don't think it's going to be here that long," said Chatters, explaining the situation with the Umatilla.

"Well, then why don't you just ship it here?" Owsley said, explaining that the Smithsonian possessed all the necessary equipment and facilities to measure and photograph Kennewick Man.

Chatters hesitated. "Well, I'm worried about it breaking."

"Oh, it'll be fine," Doug said. "We send skeletons all the time."

"I'm just not comfortable shipping it," he said.

"Well, then can you bring it here yourself?" Doug asked.

Chatters explained that he could probably get clearance from the coroner to transport Kennewick Man. But he doubted he could get funding.

"All right. Let me see if I can get approval to fly you out with the skeleton," Owsley said.

The next day he called Chatters back with a confirmed ticket for Chatters and the skeleton to fly into Washington, D.C., on September 8.

15

DESTINATION UNKNOWN

As soon as he hung up with Owsley, Chatters notified Floyd Johnson about the plane ticket. Then his doorbell rang. Grover Krantz, Chatters's former professor at Washington State University, had showed up. Days earlier, Chatters had asked him to come and see the Kennewick Man, figuring it couldn't hurt to have one more physical anthropologist examine him. Krantz spent an hour before saying he believed the skeleton could not be associated with Native Americans or any other modern populations that he was familiar with. He agreed to provide a letter verifying his opinion.

Floyd Johnson had barely arrived home from work when the phone rang. It was almost 5:00 P.M.

"Hello."

"This is Linda Kirts. I'm an attorney with the Army Corps of Engineers."

Johnson cleared his throat. "Yes."

She wanted to know why he was still involved with the remains.

"Well . . ." He hesitated, unprepared for her direct tone. "It falls within my jurisdiction to—"

"Those remains were found on federal land," Kirts interrupted.

"Well, all remains that are found in the county are under the coroner's jurisdiction," Johnson said.

"You may be violating federal law by keeping those remains, not to mention angering Native Americans." She demanded to know where Johnson was storing the remains.

"Well, they are in Dr. Chatters's office," Johnson said.

"Are they *secure?*"

"Yes, they're secure."

"Well, I think they should be locked up somewhere other than in Dr. Chatters's office."

"They're being cared for in a proper manner."

"Just what are you planning to do with the remains?"

"Dr. Chatters has a plane ticket and is preparing to take the remains to the Smithsonian."

There was a long pause, causing Johnson to wonder if Kirts had hung up.

"Hello?" he said.

"You better rethink this process," she finally said, informing him that he had no right to send the remains to the Smithsonian. Kirts told Johnson that it was beyond his scope of authority to make that determination, and insisted he would be violating the law by letting Kennewick Man go to the Smithsonian.

"But these are not Native American remains," Johnson said, explaining that Chatters and several of his colleagues concurred with these findings. "And I don't believe the Native American Graves Protection and Repatriation Act is applicable here."

"I'm not talking about NAGPRA," Kirts said, raising her voice. "I'm talking about ARPA, the Archaeological Resources Protection Act." In 1979 Congress passed ARPA to protect from commercial exploitation all federal or Indian lands containing archaeologically important human remains or artifacts. Before excavation or removal of such material can take place, the federal agency over the land in question has to issue a permit. The permit that Chatters had obtained from the corps was issued under this law.

Johnson was confused. "Ms. Kirts, I'm not familiar with that law. And I don't have to talk to you. I feel I have jurisdiction. If you want, you can talk to my attorney. You can call Andy Miller, the county attorney. Good-bye, Ms. Kirts."

Hanging up, Johnson stared at the pictures hanging on the wall next to his desk. In each one, Johnson stood next to famous country singers: Johnny Cash, Crystal Gayle, Rick Skaggs, and others. The pictures were taken in the mid-1980s, when Johnson was still a police officer. Providing security to famous singers at the annual county fair was the closest Johnson had ever been to the spotlight. He had a sinking feeling that was about to change.

He quickly called Andy Miller, hoping to get to him before Kirts did.

"Hey, Andy. This is Floyd," he said, launching right into a blow-by-blow account of his conversation with Kirts. Unable to get a word in edgewise, Miller just listened. "Andy," Johnson continued, "she was so harsh and condescending, I just told her, 'Hey, I'm not gonna talk to you.' And I hung up on her."

Miller tried not to laugh. In all the years he had known Johnson, he had never heard him angry. It was almost comical. Who is this person that got Floyd so upset? he wondered. As soon as he hung up with Johnson, Miller found out.

"We need the remains in federal custody *right away,*" Kirts demanded, after announcing herself.

Put off by the tone and volume of her voice, Miller said nothing. Kirts complained about the remains being at Chatters's office and argued that he could not be trusted to store them in safekeeping. Up to this point, Miller, whose job was to prosecute crimes, had had no involvement with the remains and no idea where they were being stored or what their status was.

Miller listened politely as Kirts started quoting the Archaeological Resources Protection Act. Kirts told Miller that under ARPA he had no choice but to order Floyd Johnson to surrender the remains to the corps.

Miller had never heard of ARPA and did not intend to make decisions based on a law he hadn't read. "Look," said Miller, "I don't have the law in front of me."

Kirts interrupted again, this time threatening to call the U.S. attorney's office to report Miller's unwillingness to comply with federal law.

Miller took a deep breath. It was Labor Day weekend and all state and federal offices were closed until Tuesday. "You are calling me at home on a *Friday night, on Labor Day weekend,*" Miller said. "Why don't we all sit down and discuss this."

They agreed to meet first thing Wednesday morning.

"But I want those remains secured right now," Kirts demanded.

Irritated, Miller sighed. "Look, Ms. Kirts, if there's any concern, I can make sure the remains are secure. The coroner can secure them at the Benton County Sheriff's Office."

Miller hung up with Kirts and called Johnson back.

Hardly able to wait to take Kennewick Man to the Smithsonian, Chatters loaded film into his camera, preparing to make his own photographic inventory of the skeleton before the trip. The phone rang.

"Paleoscience."

"Hey, buddy, I got some bad news," Floyd Johnson said.

"What?" Chatters asked.

"I'm gonna have to come over and get the bones."

"What? Why?"

"Because that's what the Corps of Engineers and my attorney say has to be done."

"Floyd, I thought you said *you* were in charge of this until the analysis was over, and they couldn't interfere with you."

"Well, the lawyers are in it now, and the county attorney has advised me to do what the corps wants."

"But *I'm not through yet.*"

"Well, I'm supposed to come right away."

After hanging up, Chatters told his wife, Jenny, the news.

"This is bizarre," she said, noticing that her husband's hands were shaking and he was breathing irregularly. "What in the hell is the government doing?"

Frantic, Jim scurried to load film into his camera. "I haven't taken enough photographs," he said.

Jenny suggested using a videocamera too.

Before he could answer, Chatters's friend Tom McClelland came downstairs into the lab area. A sculptor, he had made a cast of the Kennewick Man's skull.

"What's going on?" McClelland asked, observing the look of panic in his friend's eyes.

"Floyd called, and they're coming to pick up the bones," Chatters said, looking into the camera.

"When are they coming?"

"Right away."

"What can I do to help?"

Jenny had located the video camera and inserted a cassette. Jim handed the camcorder to McClelland. "Here," he said. "One of the things we need to do is videotape the skeleton. That would be a big help."

The skeleton was laid out on the table. "Just pan this thing back and forth," Chatters said. "Zoom in and out on the skeleton as much as you can."

McClelland stood at the foot of the skeleton and started recording. Chatters got back behind the camera and began snapping. He had gone through five rolls of film by the time Johnson and his deputy arrived at 7:00 P.M.

"Don't panic, guys," said Johnson, observing Chatters racing around the lab. "Take your time. Go ahead and finish up what you're doing."

Chatters stopped taking pictures and started unscrewing the plywood top to the wooden box that he had planned to use for transporting Kennewick Man to the Smithsonian. Carefully, he picked up four bones that constituted the two femurs—the largest bones in the skeleton—and placed them in two oversize Ziploc storage bags. He put the femurs in the box first. Then he lifted the skull, gingerly placing it in a Ziploc bag, and set it in the bottom of the box opposite the femurs.

Johnson, his deputy, and McClelland watched silently as Chatters systematically packed each bone in plastic, then stacked them in the box.

Jenny watched her husband too, knowing he felt a sense of failure. As a scientist who had been entrusted with the Kennewick Man, he felt a responsibility to learn everything he could about him. The more Jenny watched him pack the bones to give to the government, the angrier she became. How can the government do something like this? she thought. She felt like a government agency was telling a scientist that his work was basically contraband. To her, Jim was just doing his job.

After placing the last plastic bag inside the box, Chatters screwed the cover back on. Johnson wrapped the box in yellow "Evidence" tape, forming a protective seal. Jim and his wife trailed Johnson to their front door, where they stood and watched him load Kennewick Man into his Jeep. Neither of them spoke. Kennewick Man had rested in their basement lab for a month.

A 19th century American Indian skeleton recovered from Wyoming's Pitchfork Cave in 1973. Owsley's experience in this cave was pivotal in deciding his career direction. *(Photo by George Gill)*

The Smithsonian's administrative office building, otherwise known as "The Castle." Founded in 1846, the Smithsonian is now the world's largest museum complex, with more than sixteen museums, four research centers, the National Zoo, and vast libraries. It also holds the world's largest repository of human skeletons. *(Photo by Chip Clark)*

Owsley reassembling the charred skull of David Koresh following the fire at the Branch Davidian compound in Waco, Texas, in 1993. Owsley's examination of the skull revealed that Koresh had been killed by a single gunshot to the head prior to the fire. The exit wound left behind by the bullet is visible in the top left corner of the photograph.

(Photo by Chip Clark)

In 1992, in historic St. Mary's City, Maryland, lead coffins from the 1600s were excavated from the foundation of the first Catholic church built in the colonial era. Owsley and his assistant Kari Sandness remove remains from one of the Calvert coffins.

(Photo by Chip Clark)

Owsley assisted archeologists by identifying skeletons from the burials at the original Jamestown Fort, constructed after London's Virginia Company settled in Jamestown in 1607.
(Photo by Chip Clark)

In 1996, Owsley discovered the first African skeletons recovered at Jamestown. This skull, belonging to an African adult male, contains evidence of an advanced case of syphilis and a gunshot wound.
(Photo by Chip Clark)

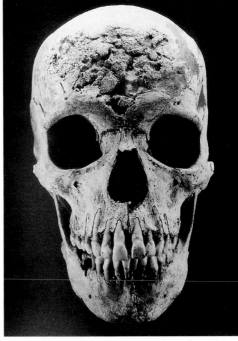

In 1996, the 9,600-year-old Kennewick Man surfaced along a riverbank of the Columbia River in Kennewick, Washington. The discovery of these bones would lead to a protracted and landmark lawsuit.
(Photo courtesy of Floyd Johnson)

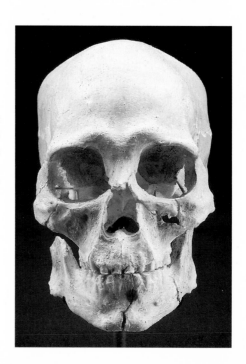

This cast of Kennewick Man's skull was made shortly before the government took possession of the skeleton.
(Photo by Chip Clark)

Within hours of the discovery of Kennewick Man, Benton County coroner Floyd Johnson examined it and concluded it was not a recent homicide victim. Johnson maintained custody of Kennewick Man until the government ordered him to surrender the remains.
(Photo courtesy of Floyd Johnson)

After being asked by the coroner's office to examine Kennewick Man, paleontologist James Chatters concluded the remains could not be Native American. Here Chatters studies the cast of Kennewick Man.
(Photo courtesy of James Chatters)

After the Army corps stopped the transfer of Kennewick Man to the Smithsonian, Owsley retained Portland attorney Alan Schneider, one of the nation's leading experts on NAGPRA.
(Photo courtesy of Alan Schneider)

Attorney Paula Barran is one of Portland, Oregon's top litigators. In 1996, she teamed up with Schneider to represent Owsley and seven other scientists in their lawsuit.
(Photo courtesy of Paula Barran)

Judge John Jelderks presided over the Kennewick Man lawsuit that began in 1996. In September 2002 he issued his decision.
(Photo courtesy of Judge Jelderks)

The chairman of the anthropology department at the Smithsonian National Museum of Natural History, Dennis Stanford joined Owsley in the legal battle over the right to study Kennewick Man. Stanford is the nation's leading expert on Clovis artifacts, pictured here. *(Photo by Chip Clark)*

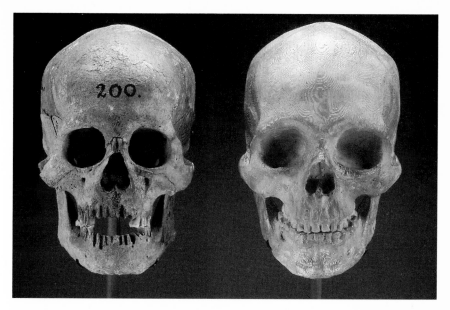

Doug Owsley and his colleague Richard Jantz studied the 10,000-year-old
Spirit Cave mummy. The mummy most closely resembles the Ainu, a
maritime people who date back 10,000 years. The skull on the right is a cast
of the Spirit Cave Mummy's skull. The skull on the left is an Ainu.
(Photo by Chip Clark)

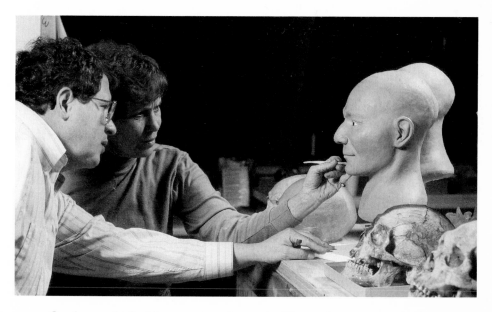

Owsley worked with reconstruction artist Sharon Long to create a bust
of the Spirit Cave Mummy. *(Photo by Chip Clark)*

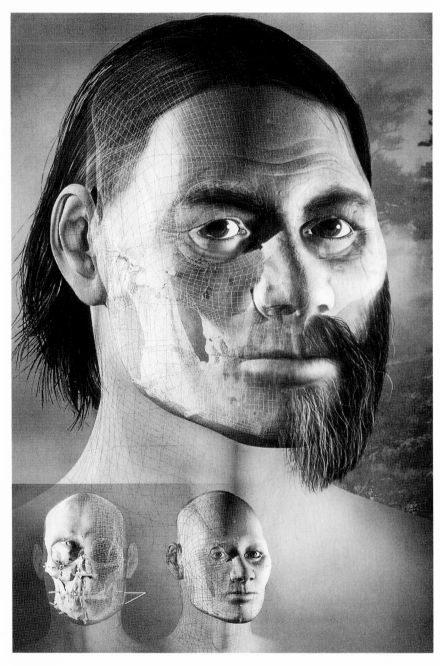

Owsley was the lead consultant on this artistic rendition of Kennewick Man.
*(*National Geographic *photo)*

16

RUSH TO REPATRIATE

It was Labor Day weekend. But Owsley could not resist working. All night he had trouble sleeping. On nights following work on cases like Waco, he could fall asleep in a heartbeat and sleep like a babe. But new discoveries of old skeletons turned sleep into an enemy. Thinking about Kennewick Man, Doug tossed and turned. When he awoke, he couldn't relax. He got in his car and headed to his office.

Susie worried about Doug's being alone in a car. He regularly wore a seat belt, drove the speed limit, and kept his eyes on the road. But his mind was never in the car. Mentally, he was always at work. Even the car radio could not distract him. Lyrics rarely penetrated his mind. Aware of Susie's fear, he actively tried to concentrate whenever he engaged in tasks like driving or operating a lawn mower. But his attempts proved futile. If alone, Doug could not stop his mind from running to his work. It was the place he felt most at home, laying out skeletons or piecing together research designs. When he could not do it with his hands, he performed the task in his thoughts.

The anticipation of Kennewick Man's arrival at his office caused him to visualize the research possibilities. He sensed that when coupled with the Spirit Cave mummy, Kennewick Man promised an even deeper view into America's ancient past, a view no scientist had ever seen.

Once at his office, he visualized the research scenarios. Jantz would take a complete set of skull measurements. Chip Clark would take high-definition photographs as well as X rays. And Owsley planned to measure all the bones and score the bone and teeth pathologies. He hoped that he would determine cause of death by examining the hip injury that Chatters had described.

He also anticipated the data he could get from the teeth, figuring they likely held clues to whether Kennewick Man had migrated to the Columbia Basin, and—if so—from what direction. Bone pathologies would show potential signs of trauma and sources of weaponry used to inflict trauma. Weaponry embedded in bone—such as the spear point in the hip—held geological clues that could help place Kennewick Man in time and space. And the dental pathologies on Kennewick Man's thirty teeth would disclose clues to diet.

With the skeleton's teeth so well preserved, Owsley figured he'd be able to detect cavities, tooth loss, abscesses, and wear. Heavy wearing on the teeth suggests a diet of coarsely processed foods like seeds and grains. A more refined wearing pattern suggests softer foods, such as fish and vegetation. By working with a network of nutritionists, botanists, fossil experts, and archaeologists, Owsley planned to reconstruct Kennewick Man's diet. From there, his geographic range could be defined by determining what regions possessed the climate to produce such foods. Eventually, he should be able to know something about Kennewick Man's migration pattern, not to mention the clues that DNA testing would contribute too. By the time they were done, Owsley felt, they would know whether Kennewick Man was the ancestor of a modern human population or whether he was from a population that today's scientists have never encountered.

Owsley quickly drafted a letter to Floyd Johnson, assuring him that Kennewick Man was a national treasure that would be properly cared for at the Smithsonian.

Wednesday, September 4
Benton County Justice Center

Army Corps Lieutenant Colonel Donald Curtis, dressed in full military uniform, trailed attorney Linda Kirts into the lobby outside prosecutor Andy Miller's office. Joined by his thirty-one-year-old deputy, Ryan Brown, Miller greeted them and turned toward the conference room.

"I have a list of people who aren't going to be here and want to be conferenced in," Kirts said, handing Miller a piece of paper. It contained the names and phone numbers for tribal leaders and lawyers from the five confederated tribes in Oregon, Washington, and Idaho, as well as a Umatilla lawyer in Colorado.

Surprised, Miller and his deputy looked at the list. "We didn't realize this was going to be a teleconference," Miller said. "I don't know if our phone system will handle this many lines."

"Well, if you can't handle this we should have had this meeting at the corps' offices," said Kirts, informing Miller that the tribes weren't the only ones who needed to be conferenced in. Bruce Didesch, the assistant U.S. attorney in Spokane, needed to be patched in too.

"We'll do our best," said Miller. "But it was your idea to have this meeting here, not ours."

Miller escorted Kirts, Curtis, and two corps archaeologists—Ray Tracy and John Leier—into the law library. They took seats at the conference table, opposite Floyd Johnson and Jim Chatters. While Miller and Brown tried patching tribal leaders in through the cream-colored speakerphone at the center of the table, the corps archaeologists stared down at the rust-colored carpet, avoiding eye contact with Johnson and Chatters.

Only able to patch in the five tribal representatives, Miller had a separate phone brought in for Lieutenant Colonel Curtis, who dialed up the U.S. attorney's office. Cupping the receiver with his hand and speaking in a hushed tone, Curtis relayed dialogue to Bruce Didesch.

"I want to go over the applicable law," Kirts began after introducing all the parties. She handed Miller a four-page document. The words DEPARTMENT OF THE ARMY: ARCHAEOLOGICAL RESOURCES PROTECTION ACT PERMIT appeared across the top. It was dated July 30, 1996. Miller had not previously seen the permit issued by the corps to Chatters, authorizing him to recover remains and artifacts along the Columbia River.

Glancing at it quickly, Miller handed it to his deputy. In preparation for the meeting, Brown had spent the previous day researching NAGPRA on the Internet. He had read the ARPA law too, but he had not researched it and had not previously seen a copy of the permit. While Brown looked the permit over, Kirts gave the tribal representatives a chance to speak.

Dan Hester, the Colorado-based lawyer representing the Umatilla, immediately criticized Chatters, paraphrasing from a letter that Washington State professor Grover Krantz had faxed to Chatters over the weekend. "This letter summarizes my findings regarding the 9,000-year-old human skeleton that I examined at your home in Richland, WA, on August 30, 1996," Krantz's letter read. "My hour-long study, from 3:45 to 4:45 that afternoon, was easily enough to satisfy me that this skeleton cannot be racially or culturally associated with any existing American Indian group."

"The letter and the facts therein outrage the Umatilla," Hester said through the speakerphone. "They're offended that Dr. Chatters had this ancestor in his house. It's sacrilegious."

Staring at the phone, Chatters bristled.

"Well, I haven't read the letter," remarked Rory Flintknife, a lawyer and member of the Yakima, who was also on speakerphone. "But I'm offended too."

Chatters cleared his throat. "For the record, my laboratory is in my home," he said. "*All* the forensic analysis I do takes place in that laboratory. And all remains are treated with due care and respect."

Glaring at Chatters, Kirts chastised him for making plans to ship the remains to the Smithsonian, insisting that that was not going to happen. "What do you want us to do with the bones?" she asked the tribes.

"The remains should be buried today," one tribal representative said. Another tribal attorney agreed.

Miller sensed the meeting slipping out of control.

"Hold it," he said, raising his voice. "This is *not* the meeting we agreed to have. We are not prepared to make a final decision and bury the remains *today*."

"Under ARPA, the remains should be turned over to the corps," said Kirts. "And these remains should be interred as soon as possible."

"In our prior discussions you had been citing NAGPRA," Miller argued. "Now you're relying on ARPA. I haven't had time to research all of these federal laws."

Kirts became indignant. "You don't understand," she said, raising her voice. "These remains were found on fee title land managed by the corps. The ARPA permit clearly applies."

"I'm not familiar with the permit and would need more time to review it," said Miller.

Suddenly, a member of the Colville tribe interrupted, and suggested storing the remains at Pacific Northwest Laboratories (Battelle), a private nonprofit facility on the nearby Hanford Reservation, until the two sides sorted out the remains' final destination.

"The ARPA permit states the remains are the property of the Corps of Engineers," chimed in Umatilla attorney Dan Hester. "And they should be transferred to Battelle."

Kirts looked at Miller. "Will you *at least* turn it over to Battelle?"

Miller looked to Johnson and Chatters for their opinion. They remained silent. Puzzled, Miller requested a brief recess and asked to speak with Johnson and Chatters privately.

They followed Miller and Brown into Miller's office and closed the door.

"Neither of you has said anything in there," Miller said, frustrated. "I want to know how far you want to take this thing. Do you guys want us to fight the transfer of the bones to Battelle?"

"No," Chatters said, convinced that a fight was futile at this stage.

A knock at the door interrupted the meeting. Lieutenant Colonel Curtis stuck his head in and looked directly at Miller. "You and Linda have really been going at each other," he said. "We need to calm down. I know how lawyers get. But with you two in the same room, we just need to lower the adversarial thing."

"Well, when your lawyer says these things, I don't know how you expect me to answer. You ought to talk to her," Miller said.

Moments later, Miller's group returned to the meeting.

"The county will agree to transfer the remains to Battelle for temporary storage," Miller informed Kirts.

"Now that you've agreed to do that," Kirts said, "we also want you to agree that you're relinquishing all ownership, control, and authority over these remains."

"Nooo," Miller said, "that's *not* what we agreed to. You asked us to put these out at Battelle 'until this gets sorted out.' And we said yes. But that doesn't mean we have agreed with your position as to ownership. We just saw the ARPA permit for the first time today."

"What is your legal authority for ownership?" Kirts demanded.

Before Miller could respond, Yakima tribal leader Rory Flintknife reiterated the idea of using a temporary storage facility and putting off the ownership issue for another day.

The meeting ended when Floyd Johnson promised to call all parties with logistics for the transfer of the remains to Battelle.

After everyone left, Miller and Brown studied ARPA and the ARPA permit more carefully. The "Conditions" section of the permit said, "Collections of archaeological resources, artifacts and other material removed from public lands under the provisions of this permit remain the property of the United States Government and may be recalled at any time for use of the Department of the Army or other agencies of the Federal Government."

The language of the permit was clear. The bones had been found on federal land and were the property of the federal government. And the permit expressly said that anything Chatters found was subject to being recalled without notice or reason. Miller and Brown agreed that the county had no basis for exerting ownership or control over the remains.

The following morning, Miller faxed Kirts a letter: "We have now had an opportunity to review the permit as well as ARPA . . . and agree that the Corps is authorized to exert control over the remains. As such, Benton County's responsibility over the remains has terminated, and the Corps may proceed to exert full control over them."

17

WE KNOW HOW TIME BEGAN

———————————

September 6

"We were blindsided," Chatters told Owsley, calling him to cancel the airline tickets.

"What's going on?" Owsley asked.

Chatters rehearsed the previous day's events.

"I've never heard of that kind of thing before," Owsley said. "You *always* let the coroner, or whoever, write his report before taking it. Man, that's weird."

"The government just muscled the skeleton away."

"Let me make some phone calls," Owsley said, confident he could reason with the corps. "Who's the archaeologist that's handling this? What's his number?"

Moments later, Owsley called corps archaeologist John Leier in Walla Walla, Washington.

"Hi, John, this is Doug Owsley from the anthropology department at the Smithsonian," he said, itching to hear what the corps planned to do with Kennewick Man.

Leier quickly made it clear that the corps had a legal obligation to return the remains to the Indians.

"This is not necessarily Indian," Owsley said, emphasizing the importance of properly documenting and identifying Kennewick Man before making any rash decisions on his final destiny.

Leier explained that the corps had assumed responsibility for Kennewick Man and had secured him in a neutral facility. He emphasized that the corps was ensuring the bones' safety, but he made no promises about the corps changing its plans to repatriate.

"Well, I'm willing to offer my assistance to the corps as a forensic anthropologist."

"I'll log in your call," Leier said before hanging up.

Owsley couldn't help but think that something was not right.

September 10
Battelle Laboratory

Peering into the wooden box resting on an examination table, Julie Longenecker stared at Ziploc bags full of bones that composed Kennewick Man. An archaeologist employed by the Umatilla tribe, Longenecker stood next to Army Corps archaeologist Ray Tracy, preparing to remove the bags and make a written inventory of the bones. When Battelle officials took custody of Kennewick Man five days earlier, they did not conduct an inventory or an assessment to check the condition of the bones.

Before beginning, Longenecker glanced around the cramped room. More than a dozen tribal members and tribal elders stared back at her. She took a deep breath. She hoped she could identify the bones. Neither she nor Tracy was a bone expert.

One by one, Tracy removed bags and numbered them. Simultaneously, Longenecker described the bags' contents on her notepad: "cranial frag," "mandible w/tooth," "15 rib blade fgs," "Innominate (Illium) w/proj. point imbedded."

After identifying sixty-seven bags with different bones or sets of bone fragments, Longenecker and Tracy reached the bottom of the box. Only the skull remained.

They stared up at the tribal members.

"Can we pull the skull out of the box?" Tracy asked.

"No!" a tribal member insisted. "We're not going to do that."

Longenecker went to her pad. "Cranium," she wrote on the last line of the inventory. "Was not taken out of box."

Tracy and Longenecker started repacking the box, placing the longer bones first, followed by smaller bones and fragments. Tribal members stood over the bones, burning cedar twigs they had brought with them. Ashes dropped in among the bones. Before Tracy put the lid on the box, tribal members placed the cedar twigs inside, scattering them across the tops of the Ziploc bags.

For the next week, Owsley placed calls to numerous ranking Army Corps officials. Each one of them referred him to someone else. Not accustomed to getting the runaround from federal agencies, Doug reached out directly to the Umatilla and sent them a letter requesting permission to examine the skeleton. He offered to share his findings with them and suggested collaboration between the tribal government and the Smithsonian in releasing the results.

Like the corps, the Umatilla did not respond. Then on September 17, the corps published an official legal notice in Kennewick's newspaper, the *Tri-City Herald*:

"Notice is hereby given under the Native American Graves Protection and Repatriation Act of the intent to repatriate human remains in possession of the U.S. Army Corps of Engineers," it began, noting that the remains had been found on federal land believed to be the aboriginal territory of the Umatilla. "There is a relationship of shared group identity which can be reasonably traced between the human remains and five (5) Columbia River basin tribes and bands.

"Representatives of any other Native American tribe which believes itself to be culturally affiliated with these human remains should contact . . . U.S. Army Corps of Engineers," the notice concluded, "before October 23, 1996. Repatriation may begin after this date if no additional claimants come forward."

Seated at his desk and looking at a faxed copy of the notice, Owsley scanned back to the paragraph about geography. The corps had said it had decided to give Kennewick Man to the Umatilla because the tribe claimed the discovery site was on its aboriginal land. Owsley set the fax down.

To assume that Kennewick Man was the direct ancestor of a tribe inhabiting that region *today*, he believed, assumed there had been no migration in or out of that area for more than 9,800 years. But with human mobility as it was generally accepted and the Kennewick Man being so old, how could they call him the direct kin of any localized tribe or world population? He was at least 450 generations removed from Owsley's contemporaries.

Owsley looked at the notice again. "There is a relationship of shared group identity which can be reasonably traced . . . ," he reread.

Frustrated, he set the notice aside. How can they say they know that? he thought. They haven't even studied it. To him, it did not seem right to reach those conclusions without doing any research.

September 18
Portland, Oregon

Less than six feet tall, slender, and with a receding hairline, attorney Alan Schneider sat in his cramped law office, listening to messages on his answering machine. One "urgent" message was left by Owsley and contained his phone number at the Smithsonian. Schneider jotted it down.

Raised in Boise, Idaho, in the 1940s and 1950s, Alan Schneider grew up in a blue-collar, *Leave It to Beaver* neighborhood. His father was a welder; his mother, one of fourteen children, worked as a law clerk to the bankruptcy referee for the state of Idaho.

Captivated by his mother's boss—a lawyer who looked and sounded like Raymond Burr, star of Schneider's favorite television program, *Perry Mason*—Schneider set his sights on emulating him. He left Idaho in the early sixties to attend the University of San Francisco, where he earned a degree in political science. In 1965 Schneider entered law school at Stanford and ended up on its prestigious *Law*

Review. When Bobby Kennedy ran for president, Schneider and other top Stanford law students conducted voting-pattern research in California for the campaign. His research ended when Kennedy was assassinated in Los Angeles, and so did Schneider's involvement with politics. He went to a law firm and practiced business law.

Fifteen years into his legal career, Schneider went solo, opening his own office in Portland in 1986. His adult hobby being archaeology, he became involved with the Oregon Archaeological Society. He started going on digs with the National Forest Service. By the mid-nineties, the Forest Service asked him to put his legal background to work for them and teach its employees how to interpret and apply NAGPRA in cases where old bones were found on federal land. Soon thereafter he sat on the board of Oregon State University's Institute for the Study of First Americans.

Professor Robson Bonnichsen directed Oregon State's institute and knew Schneider and his expertise well. When Owsley called Bonnichsen looking for a lawyer to talk to about saving Kennewick Man, Bonnichsen suggested Schneider.

"Hi, Doug, this is Alan Schneider in Portland," Schneider began when he called Owsley. He told Owsley up front that he was very familiar with the Kennewick situation. Bonnichsen and Chatters had been keeping Schneider posted. And Schneider had advised Benton County prosecutor Andy Miller to turn the skeleton over to the corps in compliance with the archaeological permit that governed it.

"What can we do to stop this skeleton from going back in the ground?" Owsley asked.

"Well, it's unfortunate that it's gone this far," Schneider said. "The first mistake the corps made is in assuming that the remains are Native American. Because the law is clear on this point—the remains must be related to a living Native American group or tribe." Schneider explained that one of the professors that Dr. Chatters had examine these remains said this population died out and the morphology of these remains shows an unusual constellation of features.

Owsley immediately detected Schneider's familiarity with anthropological terms, something he didn't expect from a lawyer.

"The other problem with the corps' decision," Schneider continued, "is that there's nothing in the legislation that says the remains can't be studied. That's just the Army Corps saying that."

Owsley liked Schneider's approach. "Alan, can we sue the government?"

"Sure. You can *sue* the government."

"Can we win?"

"Well, I think you've got a real good case here, Doug. I mean, the corps is clearly misapplying the law."

"Because I'm willing to do whatever it takes," Owsley interrupted.

"Well, Doug, I think we should try to resolve this without litigation. We don't want to go there if we don't have to."

"It's not my preference to go to court either. But this skeleton should not be turned over without being properly identified. It's too rare, too old."

Schneider agreed. He suggested a two-pronged strategy to pressure the corps and raise public awareness. He wanted Owsley to rally his colleagues in the science community and besiege the corps with letters that stressed the importance of this discovery and the need to conduct proper study and analysis before making any decision on repatriation. Schneider planned to call the lieutenant colonel overseeing Kennewick Man and put him on notice that many concerned scientists were aware of what was going on and felt that study was warranted before making any repatriation decision.

Schneider later called Owsley back with a report on his conversation with Lieutenant Colonel Curtis.

Schneider glanced down at his notes. "He said, 'Some possibility that study would be possible,'" Schneider said, chuckling. "He would 'consider it.' But he would 'not consider it or make a decision until after the notice period has passed on their notice of repatriation.'

"He also stated that they were going to 'keep an open mind.' But that 'it was for the scientific community to get the statute amended.'"

"He's not flexible at all," said Owsley.

"No. The tribes are getting the skeleton."

Owsley was silent.

"We need to step up our efforts," said Schneider, who planned to call U.S. representative Doc Hastings, whose congressional district took in the area where Kennewick Man surfaced. "I'm going to see if there's anything Congress can do on our behalf."

Owsley hung up. He made a list of his friends in the fields of anthropology and archaeology, former professors, and colleagues at the Smithsonian. He had to alert them to what was going on and enlist their help. The first place he went was down the hall to the office of his good friend Dennis Stanford, the head of the Smithsonian's anthropology department. A burly, bearded man who always wore suspenders, Stanford listened as Owsley explained what was going on.

A decade earlier, Stanford had had a key role hiring Owsley. The two of them shared a lot in common, not the least of which were their Wyoming roots and their passion for studying and helping American Indians. While interviewing Owsley for the Smithsonian curator position, Stanford revealed that he knew Owsley's father, Bill. Stanford explained that one of his best friends lived in Lusk and was a young engineer who discovered the Hell Gap archaeological site that Stanford had dug in his student days. Owsley's father was the game warden responsible for that region, and he knew the engineer that Stanford worked with.

From the day they met, Stanford and Owsley felt as if they had known each other forever. They became much closer after Stanford was so instrumental in hiring him. Stanford offered some advice and his full support. Next Owsley called Richard Jantz at Tennessee. Eager to help keep Kennewick Man aboveground, Jantz thought he and Owsley should write a joint letter. One of Jantz's graduate students then suggested a way to instantly get the letter into the hands of a vast audience: the World Wide Web. The student sent the Owsley-Jantz missive by E-mail to anthropologists and archaeologists all over the world. "We have less than 30 days before the remains are turned over," it read. "If you wish to express your displeasure with the Corps' action, please write or fax: MG Ernest J. Harrell, Commander, North Pacific Division, U.S. Army Corps of Engineers."

Faxes from around the world flooded the offices of the Army Corps, expressing shock and dismay. Schneider and Owsley's efforts paid off. By September 30, the Army Corps' decision to prevent study of Kennewick Man had reached the front page of the *New York Times*, under the headline "Tribe Stops Study of Bones That Challenge Its History." Owsley was quoted in the article about Kennewick Man's importance. "If there is no further opportunity to examine these remains, we will be losing information that is important to every American."

In the same article, the Umatilla challenged Owsley and the scientists. "Our oral history goes back 10,000 years," Armand Minthorn told the *Times*. "We know how time began and how Indian people were created. They can say whatever they want, the scientists. They are being disrespectful."

Senators and congressmen reacted to the news with dismay. On October 4, Congressman Hastings wrote to the commander of the United States Army Corps in Washington, D.C. "I was greatly alarmed by reports that the Army Corps' Walla Walla office intends to turn the skeleton over to the Umatilla tribe before it can be adequately studied and its origins conclusively determined," Hastings wrote. "I urge you to postpone action until the origins are determined conclusively or until Congress has the opportunity to review this important issue."

Days later, a coalition of congressmen and senators, led by Washington's U.S. senator Slade Gorton, wrote another letter to the Corps' commander, strongly urging him not to prematurely surrender Kennewick Man without proof of origin or identity.

The ante had been raised.

18

THE CLIENT

October 8
Portland, Oregon

"They don't intend to let a bunch of old bones get in the way." Schneider couldn't get the words out of his mind. After talking with Paul Rubenstein, a high-ranking Army Corps official in Washington, D.C., Schneider had summed up the corps' position: The likelihood of angering the Umatilla by subjecting Kennewick Man to scientific analysis wasn't worth the fight. The corps had far more financially important disputes with the Umatilla and other tribes in the Pacific Northwest over energy and the environment. Claiming irreparable harm to their hunting and fishing rights, tribes wanted some of the nation's biggest dams in Washington and Idaho dismantled, a step that would drastically impact the cost of producing and transmitting energy in the western states. Giving up Kennewick Man seemed to be an easy way to placate the tribes to some extent.

Schneider was amazed. The Army Corps of Engineers was a federal agency under the control of Congress. Yet the agency was deaf to requests from Congress to hold back on repatriation until proper study was performed. Nor had the power of the press been able to move the bureaucracy.

We're going to have to sue, thought Schneider. All the options—

talking to the corps, letter writing, congressional intervention, and direct appeal to the Umatilla—had failed. Nothing had persuaded the Army Corps to delay, much less change its mind about turning over Kennewick Man. The federal bureaucracy was simply immune to public pressure. And time was running out. The repatriation deadline was just fifteen days away. A court-ordered injunction is all that will stop the corps, Schneider concluded.

First he would have to apply for and obtain a temporary restraining order to delay the transfer of Kennewick Man before October 23, buying himself enough time to then research and prepare a legal argument that justified overturning the corps' decision to repatriate. To win a temporary restraining order, Schneider had to convince a judge that there was a substantial likelihood of proving that the corps did not have enough evidence to decide that Kennewick Man was culturally affiliated to the Umatilla.

Schneider knew that a court order was exactly what Owsley wanted him to pursue. But he worried that Owsley didn't know what he was getting himself into, and feared that Owsley might not have the stamina to withstand a lawsuit. Before calling him to discuss what to do next, Schneider sized up Owsley as a potential plaintiff.

Owsley was a scientist with no experience in civil litigation and even less familiarity with politics, Schneider reasoned. Litigation is like war. And by suing the federal government, Owsley would be fighting both a legal battle and a political one. The political fight would get dirty. Government lawyers would try to sully Doug's credentials and motives, and would likely dig into his past and his present private life.

This last aspect didn't concern Schneider, however. Owsley was one of the most respected scientists in the United States. Federal agencies from the FBI to the Park Service to the State Department had called on him to help solve crimes or identify human remains in seemingly unsolvable cases. Owsley was the man to call whenever the government faced mission impossible, and Schneider relished the prospect of the government trying to discredit a man who was so often relied upon as an expert in some of the high-profile cases it had prosecuted.

Schneider thought about Doug's private life. No problem there, either, he figured. An altar boy and an Eagle Scout as a kid, Doug seemed still to live by the creeds he had sworn to obey as a teen. He didn't even drink alcohol, except on rare occasions. Work was his only vice. He worked so many hours that he didn't have time to get into trouble.

Owsley's love for science concerned Schneider most. A lawsuit promised to crimp his passion for study. Because the Army Corps is a federal agency, it would be defended in court by the Justice Department, which has an endless string of lawyers and unlimited resources. The government's strategy would be simple: wear Doug down with endless delays that prolonged the case and bury him in paperwork that would stifle his ability to practice science. The government would try to force Doug to choose between doing what he loved and fighting for what he believed in.

Also, by making the case drag through court for years, the government would drive up the legal fees. Justice Department lawyers are paid with taxpayer dollars. Doug didn't even have sufficient funds to pay one month's worth of legal fees, let alone years' worth—which meant that Schneider would have to take the case on a contingency basis even though his previous experiences with long-lasting contingency cases had been less than positive. Each time, the client started out eager only to wear down over time and lose interest. Financially invested in the case, Schneider had no choice but to fight on alone, hoping to win a judgment in order to get compensated.

In order for Schneider to keep pace with the Justice Department, he would have to have another law firm's assistance. The workload would simply be impossible for one lawyer to manage. The firm would have to be large enough to underwrite the cost of the litigation. Attracting such a firm was further complicated by the fact that Owsley wasn't suing for money; he was suing to protect the integrity of science and to preserve truth. Schneider's only hope of compensation rested on his winning the case and having a judge order the government to pay a portion of Schneider's legal fees. The chances of winning the case rested largely on Owsley's stamina. Schneider would need him in order to prove that the government had improperly identified Kennewick Man's biological identity, an effort that would take hundreds of hours of Owsley's time.

Did Owsley have what it took to withstand a lawsuit against the government? Or would he peter out on Schneider a year or two into the case?

Schneider called Owsley, and after he explained his call to Rubenstein, Owsley was quiet. The American government's approach to the Kennewick Man discovery irked Owsley. Italian and Austrian officials had taken a totally different response after a similar set of circumstances began unfolding on September 19, 1991, when some explorers found a skeleton in the snow along the Austria-Italy border. Initially, officials figured that the remains were those of a recent accident victim who had frozen to death. Radiocarbon dating revealed that the remains were 5,300 years old. The skeleton was dubbed "Iceman," and his discovery was widely celebrated by both the Austrian and the Italian governments. Rather than rebury him, the governments supported research to learn more about Iceman's past. He is currently housed at the South Tyrol Museum of Archaeology in Bolzano, Italy.

To save Kennewick Man, a lawsuit seemed inevitable.

Yet Owsley had never sued or been sued in his life. In general he viewed lawsuits as detrimental to society. If he slipped and fell in a movie theater and broke his leg, he would either pay for the medical expenses with insurance or out of his own pocket. The idea of suing the theater for negligence would never enter his mind. Yet his latest phone call—the last of several—with Curtis had yielded nothing. And now Schneider's conversation with Rubenstein confirmed that the government had no intention of allowing Kennewick Man to be studied.

"Doug," Schneider said, waiting for him to say something, "if we're gonna stop this, litigation is the only resort. Do you want to go ahead with this?"

"I'm ready to go," Owsley snapped.

"Let's confirm that your colleagues are ready to go too."

"I've already checked with Dennis and Richard. We're ready to go.

"Alan," he continued, "if I can't get to this skeleton, with all my connections and my association with the Smithsonian, no scientist can. What chance would a graduate student or professor have in the future? I have to fight this."

Owsley had been thinking of little else lately.

If the skeleton got buried, it would be an awful precedent. If remains were found on federal land and not looked at properly, one could start losing even recent missing persons' bones on federal land. It was simply irresponsible to give up without having a proper identification done. And to lose Kennewick Man, well, it just seemed tragic.

"Well, we've tried everything else," Schneider said. "Hell, if Doc Hastings and the Congress can't stop these guys, who can?"

"That's why we've got to do this."

"Doug . . . ," Schneider began.

"What?"

"Doug, before you make this decision, you need to carefully consider whether you *really* want to do this. You need to understand this is going to require a lot of your time and demand a lot of your energy. This is a *major* decision on your part. If you get involved in a lawsuit of this nature, it's not going to be over in a month. This case is going to become your life!"

Owsley remembered being called "Scrapper" as a teen. The high school wrestling teammate who chose the name told Doug, "You go to it and stick to it." Doug never amounted to much as a wrestler. But as a scientist he prided himself on the characterization of being a scrapper.

"Alan," he said softly. "I've already talked to Richard Jantz at Tennessee and Dennis Stanford, the chair of anthropology here at the Smithsonian. Both of them are committed."

Schneider was too. He just wanted to make sure that Owsley understood what he was getting into. The time commitment would be tremendous. "Your studies and research will suffer some," Schneider said. "And you're going to take heat from some people in the academic community." Schneider explained that although Owsley was talking about suing the Army Corps, the case would quickly be framed as a battle between science and Indian tribes. The minute that happened, support from academia, the press, and Congress would dwindle. "It's something you need to be prepared for, Doug," Schneider said. "People are going to label you."

"Well, Alan," Owsley said, "someone has to take a stand here.

Everyone else is ducking and dodging, afraid to take this on. I have to do this. If not me, then who? I have no choice."

Owsley took a deep breath. "I don't need another skeleton. I've looked at thousands in my career," he said, his voice rising. "But if we don't take a stand here, the future is going to be worse in terms of our ability to ask questions of the past, worse for the next generation. When you go back far enough in time, the past belongs to the American public." Schneider agreed. "I have got to do this," Doug insisted. "I will not back down."

Schneider's adrenaline started rushing. He was hearing the right things. To him, Owsley was a fighter. If Schneider was going to the trenches, Owsley would have his back. He wouldn't find him hiding in a bunker. He seemed motivated to search for the truth and right wrongs.

With a lawsuit being unfamiliar territory, Owsley asked Schneider for instructions.

Get more scientists, representing a wide cross section of expertise, to join the lawsuit, Schneider said. More plaintiffs would bolster credibility and relieve some of the pressure on Owsley that would surely come from the academic community. And time was of the essence. Schneider had less than a week to file the case.

"So you're willing to represent us?" Doug asked.

"Yeah," Schneider said with a laugh, dreading the workload ahead.

19

SENIOR GIRL

A Starbucks coffee cup in her hand and some legal-size manila files tucked under her arm, forty-seven-year-old Paula Barran reached for the oversize brass handle on the solid mahogany door to her office. The brass nameplate on her wooden office door read MS. BARRAN. She still remembered the day, years earlier, when a janitor mounted it to the door. It seemed like a *long* time ago when the firm had debated whether her nameplate should read "M. Barran," "Mrs. Barran," or "Ms. Barran." Those were the days when women lawyers wore floppy ties and didn't know whether to go out to lunch with the secretaries or the attorneys.

Barran smirked and opened the door. She walked across the dark brown tweed carpet and dropped the files onto a granite-topped worktable. Her father had given the table to her as a gift when she made partner in 1985. At the time, she was the only woman partner in the law firm, Spears & Lubersky. In 1990, her firm took part in a merger that produced one of the biggest law firms on the West Coast, with law offices in Portland, Seattle, Los Angeles, San Francisco, Anchorage, and London, England. After the merger, Barran began sarcastically referring to herself as "Senior Girl." Despite having nearly three hundred lawyers in her new firm, Barran was one of only three women partners, none of whom had been a partner longer than she.

Eager to begin her morning coffee, Barran sat down behind her wide desk. She never ate breakfast, relying instead on a twenty-ounce cup of coffee with four shots of espresso to start every day—the more caffeine, the better. She knew it was addictive, but it was a tradeoff she was willing to make in exchange for the energy that helped her through the brutal schedule she imposed on herself. Most days she did not find time to read the morning newspaper until bedtime, around midnight. The only diploma she displayed on the wall behind her desk typified her tendency to overload her schedule. It signified Barran's completion of the executive M.B.A. program at the University of Oregon, which she accomplished while practicing law full-time.

Barran's other degrees—a master's degree in modern languages from Cornell and a Ph.D. in medieval Germanic literature from the University of British Columbia—were not on the wall. Those were part of her past life, before she decided to attend law school in 1977. Well on her way to becoming a college professor, Barran saw an episode of the 1976 TV series *Kate McShane*. In the episode, McShane, a lawyer played by Anne Meara, delivered an impassioned speech about law school. That sounds cool, Barran thought at the time. I want to feel that way.

Far more educated than most lawyers coming out of law school—and equally as shrewd and sophisticated—Barran quickly established herself as a tough trial lawyer in the area of labor law, specializing in cases involving employment discrimination. By the time she made partner, she specialized in defending large companies and corporations on the West Coast that were being sued—usually by female employees—for sexual harassment. It was not exactly what Barran had had in mind when she came out of law school. But she possessed a deft ability to sort out a genuine case of sexual harassment from an inflated one. She also had plenty of personal experience. She had seen women lawyers receive substantially lower salaries than their male counterparts, be kept off high-profile cases on account of their gender, and thrown out of court in humiliation by judges who did not like seeing women wearing pantsuits. She was no fan of sexist jokes. Nor was she sympathetic toward employees—male or female—who sued an employer with deep pockets because a fellow employee told a bad joke.

Her coffee cup nearly drained, Barran began responding to the stack of phone messages neatly arranged on the glass tabletop of her desk. As she left a series of return voice-mail messages, she contemplated the poster-size photograph hanging directly across from her desk. Shot from aboard one of the Apollo space missions, the photograph depicted a captivating view of the earth. It helped her keep things in perspective.

An avid art collector, Barran kept other items in the office, including a framed picture of the remnants of a library in Coventry, England. German bomber planes had reduced the library to rubble during World War II. But standing amid the destruction, a group of Englishmen were reading books. Barran liked the picture for what it said—that even when the world is in chaos around us, certain things endure. Next to it, a framed *New Yorker* cartoon illustrated two men talking over drinks. The caption had one man saying to the other, "These days, I suppose it doesn't matter whether you've read Thomas Mann or not." Below the frame, Barran kept a large, custom-made ceramic plate on top of a wooden pedestal.

She put down the phone, and it immediately rang.

It was Alan Schneider.

"Hey, Paula, would you like to sue the government with me? It involves a really big issue."

"Well . . ." She hesitated. Barran hardly knew Schneider and was unsure if he was serious. On three or four occasions they had referred clients to each other. But they had never met face-to-face. And their clients were involved in labor disputes, none of which ever involved suing the federal government. "What's the issue?"

Schneider filled her in on the discovery of Kennewick Man and that it had been carbon-dated at over 9,800 years old. "This is the Rosetta stone for scientists," he said.

Barran listened but said nothing. The number—9,800 years—was so immense that she did not know what it meant.

Schneider got her up to speed on the situation and the need to act quickly.

•   •   •

Intrigued, Barran asked him for more background on what law applied.

Schneider asked her if she was familiar with NAGPRA.

She had never heard of it.

Schneider summarized it.

On November 15, 1990, Congress passed NAGPRA in response to complaints that some museums possessed Native American skeletons. Native Americans wanted the remains returned for proper burial. As a result, museums across the country were required to inventory their skeletal collections to identify and return all American Indian remains to the appropriate tribes.

Although the law's primary focus was Indian remains in museums, it also contained a section dealing with new excavations and discoveries. Essentially, it gave Indian tribes the right to claim and take custody of any newly discovered Native American remains on federal or tribal lands. The law applied laid out a protocol:

If the remains have lineal descendants, they are returned to the lineal descendants.

If there are no lineal descendants and the remains are found on tribal land, then the remains go to the tribe living where the discovery was made.

If there are no lineal descendants and the remains are not found on tribal land, but are found on federal land, then the remains go to whatever tribe can prove the closest cultural affiliation.

If none of these scenarios apply, the last resort is for a court to determine which tribe had aboriginal occupation of the land where the remains were found.

"So what is the legal issue going to be in this case?" Barran asked.

Schneider identified two. The first was whether NAGPRA prohibited study of a 9,800-year-old skeleton. All three scientists who had examined the skull had independently determined that Kennewick Man could not be of Native American descent. And the government had inadequate evidence to say the skeleton was Native American and whether it was culturally affiliated with *any* tribe. The second issue was whether scientists—in this case, scientists for the Smithsonian Institution—had any rights to study Kennewick Man.

"The question," Schneider explained, "is, doesn't the government have to at least weigh the scientists' interest in the balance before turning the skeleton over to the tribes?"

Barran had never heard Schneider talk so fast, or with such fervor. He was nearly out of breath, yet showing no signs of slowing down. "The Smithsonian was in the process of transferring the skeleton to Washington when the Army Corps stepped in and said, 'It ain't going anywhere.' The government is ignoring a federal law that entitles the Smithsonian to take possession of ancient discoveries—ones which aren't Native American—for purposes of scientific research."

As Schneider reeled off the legal issues involved, Barran zeroed in on the political ones. Her firm was fairly conservative; if it took this case, would it be heading for a political buzz saw? Did she want to challenge an Indian tribe's claim to human remains? And if she helped sue the government, which part of the government would that be? The army? One of her partners had served as judge advocate general (JAG), the senior legal officer and chief legal adviser to the army, navy, and air force. Was there a conflict of interest if her firm took a case against the Army Corps of Engineers?

"So are you interested?" Schneider asked.

Barran paused. "Yes," she said hesitantly, unsure of what she was committing to.

"Paula, one other thing: These guys don't have any money to pay us. We'd have to take this on a contingency basis." He assured her that the scientists would see the litigation through.

"Let me check it out and see if I can get authorization from my firm to do it," she said.

Barran hung up the phone and sat motionless. She could not believe the government was potentially sitting on one of the most important discoveries in North American history. The public, she felt, had a right to know about it. And a lawsuit on behalf of Smithsonian scientists against the federal government would test the boundaries between science and politics. It might also lead to answers about the origins of the human species in America.

She got up from her chair, her mind racing between questions about the First Amendment and questions about anthropology. Then it struck her—she was a labor and employment lawyer, not a First Amendment lawyer. Nor did she have a scientific background. And she had never sued the federal government before. Reason told her to stay away from this one.

But then she told herself that it was every lawyer's dream to be involved in a case like this. Opportunities like this were the reason she had gone to law school.

Barran took the elevator down two floors to the sixth floor. Briskly she walked toward the office of Edwin A. Harnden, the busiest man in the entire firm. In addition to managing 280 lawyers, Harnden was the president of the Oregon State bar and a life fellow of the American Bar Association. Due to his harrowing schedule, Harnden was rarely at his desk, often being required to fly to any one of the firm's offices in other cities.

Noticing the door was ajar, Barran stuck her head in. "Are you here?" she asked, unable to see over the high back of his leather chair. Harnden was one of the shortest individuals in the entire firm.

"Yes!" he announced cheerfully. Wearing a conservative blue business suit with a white shirt and gold cuff links, he sprang to his feet and came out from behind his impeccably tidy desk. His wavy black hair was neatly combed. "Paula, have a seat," he said, flashing his genuine smile. Barran was always amazed at Harnden's relentlessly cheerful optimism. She was convinced he was incapable of feeling gloomy.

"Would you object to my suing the United States government on behalf of a group of scientists who want to study an ancient skeleton?" she said, staring directly at Harnden. Normally, Barran was not required to go to Harnden for permission before taking on a new client. As a partner, she was free to pick and choose her cases. But this wasn't a normal client.

She spent less than ten minutes filling him in on the case.

"OK, I'll sign off on that," Harnden said, not bothering to ask a single question. Nor did he suggest discussing it with other partners in the firm. He did not even ask for additional time to consider the matter. Barran was a little surprised.

"If we take the case, it is not going to pay its freight in cost," said Barran, convinced that she needed to explain more. "We're not going to get paid anything for this one. Although it is certainly going to be a lot of exposure for the firm."

"Paula, let's do it," Harnden said. Harnden had been nominated as the firm's managing partner on account of his prudent judgment and exceptional business-management instincts. He was not prone to rash decisions when it came to committing the firm's financial resources. Yet, after hearing only scant bits of information, he had agreed to take on a case that was far outside the firm's area of expertise and was guaranteed to be grossly expensive and extremely time consuming. He had known Barran for twenty years and had tried numerous cases with her. He had complete faith in her judgment and respect for her abilities.

Deciding whether or not to give Barran the green light to represent scientists without charging legal fees was an easy decision for Harnden. The minute he heard that the government was restricting its own scientists—ones the government presumably hired because they were the best in the country—from a serious investigative study, Harnden knew the case was important.

Barran rushed back to her office, eager to call Schneider and give him the news. As she punched in his telephone number, Barran's husband, Richard Hunt, ducked his head inside her office. "Come in, come in," she whispered, motioning him in with her left hand while holding the phone with her right hand. "This will take just a minute."

Hunt sat down in a soft leather chair and stretched out his long legs. His long torso and flat stomach made him appear even taller than his six-foot, six-inch height. The hair atop his receding hairline was not nearly as thick as his beard, a healthy mix of black and sophisticated gray. His gold-rimmed spectacles contrasted with his tanned face. The left pocket on his starched white shirt had the initials RCH embroidered on it. Wearing a chestnut tweed blazer, he looked like a distinguished Oxford professor. But Hunt was Paula's law partner. Both previously married and divorced, the two of them had met through the firm and married in 1987.

"Alan, hi, this is Paula. We decided it is a go."

"Great," Schneider said. "That's great."

"Listen, we need to get together right away to draft our complaint. We need more facts. We need you to spell out for us our justification for needing a temporary restraining order."

Hunt raised his eyebrows. The phrase "temporary restraining order" intrigued him. Hunt had more experience with obtaining temporary restraining orders than anyone in the firm. His specialty was representing large companies in trade-secrets cases. Whenever one of Hunt's business clients accused a competitor of stealing trade secrets, Hunt's job was to rush into court and secure an injunction ordering the competitor to halt immediately until a full-fledged court proceeding could be scheduled to determine whether in fact trade secrets had been stolen. But Barran had never applied for an injunction, and Hunt could not imagine why she needed one.

As Barran hung up with Schneider, Hunt's face slowly produced an inquisitive smile. "Paula, what are you getting involved in now?" he asked.

She told him.

"Why are you doing *that?*" he asked.

"This is something that is too important not to do."

"Are we ever going to get paid for it?" he asked, grinning.

Barran grinned back. She didn't know about legal fees. She was caught up in the principles at stake in the case. "How dare the United States government lock up the books?" she said. "How dare they hold back knowledge?"

"There are no anthropologists here," Hunt said. "None of us understands the significance of nine thousand eight hundred years old and what that means."

"That's why we need to take this case," she said. "If we don't, then no one will ever know what a Caucasoid male was doing in North America nine thousand eight hundred years ago."

Hunt paused. "I suppose *somebody* has got to take this case," he said.

As much as the idea offended his practical reasoning, Hunt saw the importance of the case. "Did I hear you say you were going to go for a temporary injunction?"

"Yes. We've got less than two weeks before the army turns the skeleton over." If that happened, she explained, the tribes would bury it on Indian land and the scientists would never recover it. Barran wanted an injunction from the court to delay the transfer long enough to give her and Schneider time to prepare a case and go to court.

"So you're the expert on injunctions," she said. "What do I need for an injunction?"

"Well, you don't just go file your complaint, your motion, and your brief," Hunt said. "There are all these hoops you have to jump through to get an injunction."

"What are they?"

Hunt explained.

First Barran needed to secure a hearing date at the federal court, and she could only hope that a judge would be available to hear the case on such short notice. In order to schedule a date, she had to draft a petition and get it in the hands of a court clerk at once. But even before taking the petition to a clerk, she needed to try to anticipate what the court would do, because along with the petition she needed to prepare orders for the judge to sign. If the judge granted the petition, he might ask for the orders to be revised and resubmitted. But a draft was essential.

There was another problem. As the ones filing the petition, Barran and Schneider were responsible for getting word out to all the parties they hoped to enjoin. If the judge issued an injunction, they would have to serve notice to everyone who had custody of the skeleton. Barran and Schneider would have less than twenty-four hours to do it. If just one of the various government agencies or officials involved in the chain of command did not get properly served with notice of the injunction, Barran and Schneider ran the risk of the skeleton's being transferred despite the court-ordered injunction.

"Richard, I don't even know who we are suing in this case. I mean the Army Corps has custody of the skeleton. But whom does the Army Corps answer to? Aren't they under the direction of the Interior Department? Whose name do we put on the lawsuit?"

Hunt smiled. "You have a way of these crazy situations just coming to you—or you go to them."

20

EIGHT MEN OUT

October 13, 1996
Laramie, Wyoming

When trying to determine who among the nation's anthropologists would have the courage to join him in suing the federal government, Owsley thought first of his first anthropology professor, George Gill. Still teaching anthropology at the University of Wyoming, Gill had bought a small ranch outside Laramie and built himself a log home, complete with wide-open spaces and horses. He increasingly spent time relaxing on his ranch and working out of a home office and library. He enjoyed random calls from Owsley to hear about his latest adventures at the Smithsonian.

This time, however, Owsley called looking for help. Owsley's explanation about Kennewick Man reminded Gill of the time he took Owsley up to Pitchfork Cave to recover two Indian skeletons. The rules and politics of anthropology and archaeology were much different then. The Pitchfork remains were recovered, preserved, studied, and identified without exposure to the vast federal bureaucracy created by NAGPRA. Gill ultimately concluded that both Pitchfork skeletons were Native American. Tribal affiliation was not as clear, however. But based on associated artifacts and geography, he thought they were likely Crow Indians, a nomadic tribe that inhabited northern Wyoming around the time the Pitchfork individuals died.

Gill's complete forensic report and conclusions were provided to the Crow, along with a request for them to take custody of the remains. The Crow, who now reside on a reservation in Montana, studied the report. But after consulting the tribal elders and reviewing oral traditions, the Crow politely rejected the University of Wyoming's attempt to repatriate the remains. Without more conclusive evidence, the Crow had no desire to rebury remains that might belong to another tribe, particularly one that might formerly have been an enemy.

To Gill, the government's rush to repatriate Kennewick Man seemed driven far more by politics than honoring the dead. That, he thought, would ultimately do a disservice to both Indians and scientists.

Gill agreed to join the plaintiffs.

October 14, 1996
Portland, Oregon

It was early on a rainy afternoon when Alan Schneider got off the elevator on the eighth floor of the Pacific Building. Facing a wall of mirrored panels, he turned right and pushed open a set of thick, heavy doors leading into the law firm Spears & Lubersky. Entering onto a dark hardwood parquet floor, he approached the receptionist's dark mahogany desk, behind which stood large plants in oversize black pots set in front of windows that overlooked the courthouse across the street.

Schneider introduced himself.

"Yes, right this way, Mr. Schneider."

Following the receptionist down a long, darkly carpeted corridor, Schneider clutched the soft handle of his badly worn leather bag. Its tanned rawhide exterior, weather-beaten and faded, looked more like the seat of an old horse saddle than a lawyer's briefcase. Inside it, Schneider carried draft copies of a complaint, a motion for a temporary restraining order, and a memorandum to the court supporting the motion. Barran and her associates had prepared the documents over the weekend and had copies delivered to Schneider's office for review.

Despite being impressed with the filings, Schneider could not stop fretting about the looming filing deadline. He and Paula had just five days to convince a federal judge to block the government's decision. They had to finalize their complaint, their motion for a temporary restraining order, and their memorandum to accompany the motion. Plus they had to schedule a hearing date and make oral arguments at the hearing—all before the following Wednesday. He had no idea how they would manage.

"Hi, Alan," Barran called out from the doorway of a glassed-in conference room that was larger than Schneider's entire law office. "Come in. Let me introduce you to the team."

Schneider entered and looked around. Twelve lawyers and paralegals were assembled around a shiny conference table. A smile swept across Schneider's face. Twelve lawyers! Instantly the deadline did not seem as intimidating.

For the next five hours, Schneider, Barran, and her associates refined the complaint and the motion for a temporary restraining order. "What are we really after with this lawsuit?" one of the attorneys said.

"It's not just to study Kennewick Man," Schneider said. "But it is to set a precedent with respect to the handling of all ancient remains."

Barran agreed, suggesting that the case would test their discipline as lawyers. "We need to make sure that we don't expose the plaintiffs and jeopardize their scientific credentials by making statements that can't be supported scientifically."

Lawyers are trained to advocate vigorously in advancing a client's interests, a practice that easily lends itself to lawyers' making exaggerated statements that aren't necessarily supported by evidence.

"We can't ever say the skeleton is not Native American," Schneider said. "We have to say, 'We don't know. But you can't determine that without study.'"

"Nor can we say that study will determine that it is Native American," Barran pointed out. "Because studies might be inconclusive. But we can say that you can't make an accurate determination without thorough study."

When Schneider and Barran finished meeting, Schneider called Owsley at the Smithsonian.

"Doug, I'm faxing you the complaint," he said. Schneider wanted him to review it for accuracy and make any necessary changes or modifications.

An hour later, the cover sheet of the complaint reached Owsley's office. Above the word "PLAINTIFFS" he spotted his name, along with Robson Bonnichsen, Loring Brace, George Gill, Vance Haynes, Richard L. Jantz, Dennis Stanford, and Gentry Steele.

The words "United States of America and the Department of the Army" appeared next to "DEFENDANTS."

Owlsey had never sued anyone. As he flipped past the cover sheet of the complaint, the headings seemed foreign to him: "Nature of Action," "Jurisdiction," "Venue," and "Parties." He carefully reviewed it, taking notes as he went along.

Owsley called Schneider back and suggested a few minor changes.

Schneider agreed to make them. He had one more question: "Now, Doug, are you still willing to put your name on the complaint and be named as a plaintiff?"

"You bet."

"All right."

"This is a fight that needs to be had. Let's go do it."

The next day, October 16, 1996, Barran and Schneider filed their lawsuit at the U.S. District Court in Portland, Oregon.

21

ABOUT-FACE

Friday, October 18
National Museum of Natural History
Washington, D.C.

"This is bullshit!" Dennis Stanford said to himself, banging down the receiver. Lauryn Grant, a lawyer in the Smithsonian's general counsel's office, had just informed him that the Justice Department vehemently opposed Owsley's and Stanford's participation in the lawsuit. The Justice Department, citing federal conflict-of-interest statutes, insisted that it was against the law for Stanford and Owsley, as federal employees, to sue the Army Corps. Unless they voluntarily withdrew from the suit, the Justice Department planned to bring charges against them.

Stanford telephoned Schneider immediately and told him the news.

Stunned, Schneider started laughing. "I'm gonna be real pissed off if the Smithsonian takes us off this case," Stanford said. "That will leave the other scientists hanging in the wind—alone." The reason the other scientists signed on to the lawsuit was largely due to Owsley and Stanford. "Just a minute here," Schneider said, still laughing. "The Smithsonian is telling you to withdraw from a lawsuit when the only purpose of the lawsuit is to study a new discovery that's very important for scientific reasons?"

After Stanford explained Grant's phone call, Schneider was per-plexed.

"I can't conceive that what they're saying is true," Schneider said. "They're telling you that you can't exercise your civil right in order to get access to that skeleton. How are they going to explain that to the world?"

Stanford asked Schneider to investigate whether the information that the Justice Department had given Lauryn Grant was correct.

Schneider agreed to look at the statute. But if what the Justice Department was saying was correct, that would mean that if an employee is wrongfully terminated by a government agency or if an employee is injured by another government agency, he's lost all of his civil rights. Schneider couldn't conceive of that. "Dennis, what specific conflict-of-interest law are they saying that you'll be breaking if you sue?"

"I don't know," he said.

"Well, let's find out. Go back to Lauryn." Schneider advised Stan-ford to avoid being aggressive and to get the Smithsonian to explain a couple of things, one: what they want you to do; two, why they're mak-ing this demand. Three, what's their legal basis for saying you can't par-ticipate in this lawsuit? And four, what are the consequences if you con-tinue with the lawsuit? "Let's get them to confirm all of that in writing," Schneider said.

"All right."

Schneider laughed again, though he hardly thought the situation was funny. "I mean, this would be a tremendous public relations night-mare for the Smithsonian if two of its top scientists are forced to with-draw from a case that is designed to protect and advance the interests of science."

"Tell me about it."

After hanging up with Schneider, Stanford stormed down to Owsley's office and informed him that they were being ordered to withdraw from the lawsuit.

"Can they do that?" Owsley asked.

Stanford told him what Schneider had said about the importance of getting the demand in writing and instructed Owsley to write a letter to Grant at once.

"Dennis, we've got to hang in there."

"Oh, I intend to."

When Congress drafted legislation to create the Smithsonian in 1846, John Quincy Adams envisioned a federal institution of scientists dedicated to enlightening the United States through discovery and study. "It is by this attribute," Adams wrote in the Smithsonian legislation, "that man discovers his own nature as the link between earth and heaven; as the partaker of an immortal spirit; as created for higher and more durable ends, than the countless tribes of beings which people the earth, the ocean, and the air alternately instinct with life, and melting into vapor, or moldering into dust."

Kennewick Man, perhaps more than any discovery on American soil, promised to answer questions about America's first inhabitants. Yet the Justice Department was attempting to stifle Smithsonian scientists by threatening to prosecute them. And the Smithsonian's lawyers were not siding with its scientists.

After spending the entire afternoon working on his memo to Lauryn Grant and thinking about the ramifications of being forced to withdraw, Owsley left his office for home. The minute he walked through the front door, Susie knew he had something weighing on him.

"Hi, Doug."

"You know what they're doing, Susie?"

"Who?" she asked.

"The lawyers at the Smithsonian," he said rapidly. "They are telling me that as a scientist for the Smithsonian I can't sue the Army Corps. Well, as a citizen of the United States I can."

"Why won't they back you up?"

"Public image. It doesn't look good going against the Indians."

"How's it going to look if they force you off?"

"Well, that's just it. The thing that makes me so mad is that we're not against the Indians. We're just trying to ensure that this skeleton is properly identified."

As Doug walked through the house, arguing for why he should be allowed to study Kennewick Man, Susie listened. "Well, I'm not backing down," he insisted. "This is too important."

Susie paused, then said tentatively, "Doug, this is your job. You're talking about suing the people you work for."

"Well, I should be able to do that as an individual scientist."

"But, Doug, what if you lose your job over this? What about everything you've worked so hard for? You could lose all of that."

"Susie, I have to do this."

"What about the girls, Doug? They're both going to be starting college within the next couple years. If you lose your job at the Smithsonian, how will we pay for their education? What will we do?"

"Susie, what the government's doing here isn't right. They shouldn't be blocking study of Kennewick Man. And they shouldn't be stopping Dennis and me from going to court. I can't walk away now. This is about principle. It's about right and wrong."

Susie recognized that tone of determination in Doug's voice. It was Guatemala all over again. That time he put his zeal for discovery and adventure ahead of his life. This time he was risking the job that enabled him to pursue discovery and adventure.

"You need this job, Doug," she said softly. "But this *is* important. And you may as well work for someone who believes in what you're doing and who will back you."

Doug said nothing.

"Maybe they can tell you that you can't sue as a Smithsonian scientist," Susie said. "But as a private citizen you still have rights."

Susie was with him. "If I can just hang on until the case gets argued in court," he said, "I think I'll be able to stay on as a plaintiff. I just have to hang on until next Wednesday."

Alan Schneider didn't believe the Smithsonian really wanted Owsley and Stanford to withdraw. "Give the Smithsonian some time to think about what they would really like to do," he told Owsley. It had been his experience, particularly in business practice, that a lot of times decisions are made because people have done things hastily, without thinking through the consequences and implications. "If you can sort of keep things moving," Schneider said, "but don't put people in a position where they've got to make a final decision, when they reflect on it they'll mellow a little bit."

Owsley decided he was going to make himself as scarce as possible for the next few days.

"Where will you be?" Schneider asked.

"Well, I'm supposed to go to Jamestown," he said. The Parks Service had been after him to look at some skeletons for them. This was a perfect time to go do it. As long as he wasn't around his office, Owsley figured, the Smithsonian couldn't get to him to coerce him into dropping out of the suit.

22

WHERE DID YOU GET
THESE AFRICANS?

A National Park Service ranger motioned Owsley and his assistant Kari Sandness through the entrance gate to Jamestown Island. They walked past the Captain John Smith monument and the reconstructed seventeenth-century church, and stopped to examine the excavation underway at the old Jamestown Fort.

In the early 1990s, a historical preservation organization that owns over twenty-two acres at the western edge of Jamestown Island hired archaeologist William Kelso to conduct digs. When Kelso began, only one aboveground remnant—a brick church—stood from the seventeenth-century town. But in 1995 Kelso uncovered signs of the original Jamestown Fort (so named to honor England's King James), which was the first permanent English settlement in America. Historical records confirmed that on May 13, 1607, more than one hundred men and boys backed by London's Virginia Company settled Jamestown Island. By 1610 a palisade of planks enclosed a fort, which had been strengthened with five bulwarks, each containing a watchtower.

With no evidence of it existing aboveground, historians had long believed that the original Jamestown Fort had eroded into the James River. Kelso's excavation exposed sections of the original walls, one bulwark that held a watchtower, a timber building, and more than 250,000 artifacts from the early 1600s. Inside the fort, Kelso found military armor—a helmet, a breastplate, gun parts, sword and dagger parts, powder cartridges, and ammunition ranging from small shot to cannonballs—and coins that predated 1603. He also found two graves, each one containing a skeleton. Kelso invited Owsley to help him excavate the burials and identify the remains. In September, Owsley first visited the site, examined one of the skeletons, and confirmed it was a European colonist who was between seventeen and twenty-two at the time of his death.

On this return visit, Owsley walked to the nearby Parks Service building and met David Riggs, the museum curator. Riggs had the responsibility of sorting out the skeletons recovered from Jamestown Island, separating the English colonists from the Indians. All Indian skeletons had to be repatriated under the NAGPRA law. The English colonists did not. Riggs wanted Owsley to assist him in identifying the skeletons.

Riggs escorted Owsley into the curation room, lined with stacked metal cabinets. Riggs pulled their drawers out one by one, revealing skeletons excavated from the Jamestown graveyard in the 1940s. In 1955 the National Parks Service had excavated a second cemetery on the island, as well as some individual burials that were not included in either of the two graveyards. All of the remains were curated at the Jamestown Museum.

Owsley looked carefully through each drawer, perusing the skeletons. Suddenly he stopped. "Hmm," he said, picking up a skull to examine it more closely. It had a worm-eaten appearance. "Where'd you get this African?"

Squinting, Riggs cocked his head. The National Parks Service had no record of any African skeletons being excavated at Jamestown, only European colonists and Indians.

"This isn't an Indian," Owsley said. "This is an African."

Riggs had photographs and paperwork on file about the skeleton. When discovered by archaeologists, the skeleton was positioned on its right side, its knees and elbows flexed and its hands clasped together and resting under its right cheek. It was remarkably well preserved.

Owsley examined the skeleton more closely.

Disease, thought Owsley, examining the limbs. The arms and legs are riddled with disease.

He read the identification card in the drawer. "Considering the important part that the Indians of the Powhattan Confederacy had played in the history of Jamestown and the fact that this Indian was the first Indian burial to be found in the site . . . it should be incorporated in the exhibits at the Jamestown Museum." The description had been written in 1942 and was based on archaeologists' observations that the skeleton had shovel-shaped incisors and had not been buried in a Christian manner, both indications that it was not European but, they believed, Indian.

"This is a very important skeleton," Owsley said. "We definitely need to document this collection."

Owsley promised to return.

23

TURNING THE LIGHTS ON

When Owsley returned to his office, an envelope bearing the word CONFIDENTIAL awaited him. Inside he found a memorandum from Smithsonian lawyer Lauryn Grant.

"On Monday (the 21st) we received a telephone call from the Department of Justice which is representing the interests of the Army Corps of Engineers in this litigation," the memo said. "The person who called raised several questions about the independent litigation authority of the Smithsonian. We also received a call from Steven Carroll, an attorney in the Environmental Section of the Department of Justice who is supervising this matter. He raised a question about the authority of Smithsonian federal employees to sue other federal employees, and he expressed serious concern about your participation in this lawsuit. Specifically, he asked for assurances that neither of you would be testifying tomorrow at the hearing on the temporary restraining order in Portland. If you had planned to testify, he stated that the Department of Justice would likely move to prevent your testimony as potentially adverse to the interests of the United States."

Feeling as though he was being threatened, Owsley paused before reading on.

"As Smithsonian employees, you are bound by the Standards of Conduct," the memo continued. "The Standards of Conduct require that you refrain from any private or personal activity which might conflict, or appear to conflict, with the interests of the Institution." A copy of the Institution's Standards of Conduct was attached to the memo.

"An additional concern, is the existence of criminal conflict of interest statutes which prohibit employees of the United States from acting as agent or attorney for prosecuting any claim against the United States (Title 18 U.S.C. Section 205)," the memo continued. "Although it is not clear whether, and to what extent, this provision applies, the Department of Justice has raised this issue. They strongly oppose the presence of Smithsonian employees on the opposite side."

Owsley glanced at the last paragraph. "For all these reasons, we have advised you to withdraw as plaintiffs from this litigation." He folded up the letter and put it back in the envelope. Then he called Dennis Stanford and left a message for him to stop by his office right away.

An hour later Stanford arrived.

"Doug, I'm on my way to the airport. I got a message that you needed to see me right away."

Owsley handed him the envelope. "Did you get one of these?"

Stanford put his bag down and opened the envelope. "No," he said, starting to read. Reaching the end, he started shaking his head. "This is a pretty crummy situation," he said. "It's going to be real embarrassing if we have to call these other guys back and tell them we have to pull out. We're the reason these other guys got involved in the lawsuit."

Owsley agreed.

"While I'm away, take this to Fri," Stanford said, referring to his boss, Robert Fri, the director of the Smithsonian's National Museum of Natural History. "We need his opinion. But from the looks of this, if we don't withdraw we each get a pink slip."

"It's starting to sound that way."

"My reading of it is that the Justice Department is telling the institution what we can and can't do. And there's a real possibility that if we stay in and defy the institution we might lose our jobs."

Owsley agreed. But the court proceedings were due to begin in a few hours on the West Coast. Owsley believed that once the gavel came down, it would be much more difficult for the government to force them to pull out. The adverse publicity would be too steep, particularly for the Smithsonian.

"Well, Dennis, let's hang in there until we hear from Alan."

Stanford handed the memo back to Owsley. "I'll be back in a week and a half. We'll meet then to discuss this further. And call Alan later to find out what happens in court today."

Hours later
Portland, Oregon

His black robe cloaking his tall, moderately built frame, his dark hair and mustache neatly trimmed, Judge John Jelderks entered the courtroom. A former lieutenant commander in the U.S. Naval Reserve, Jelderks became a federal magistrate in 1991, after spending nearly twenty years as a circuit judge in Oregon.

Barran, who had appeared before him numerous times on other cases, glanced at the clock on the wall: 1:39.

Jelderks had scheduled the hearing upon Barran and Schneider's request for an emergency injunction to prevent the Army Corps from transferring Kennewick Man to the tribes. The transfer was due to take place in twenty-four hours. But when Jelderks arrived at his chambers earlier in the morning he found a written notice from the Justice Department saying the transfer would be delayed. He planned to get an update from both sides on the status of the case. Then he would determine whether to issue the injunction.

He looked to Barran first and asked if, given the government's promise not to turn the skeleton over to the tribes the following day, he still needed to issue the injunction.

"We do believe that there is a need for immediate court action," Barran began. Her clients had been stiff-armed by the Army Corps and threatened by the Justice Department. As far as she was concerned, the government could not be entrusted with Kennewick Man's care without court supervision. "It strikes me that government only works when it's not done in secret," she said. "And our plaintiffs very much regret that they had to file a lawsuit to turn the lights on in this case."

Barran argued that despite what the Justice Department was telling the court, the legal notice filed in the newspaper by the Army Corps made it quite clear that the government had already made up its mind to give Kennewick Man to the tribes. Barran's position was that the corps had no legal authority under NAGPRA to repatriate Kennewick Man, since the statute applied only to Native American remains. And in her view, the government had no evidence that Kennewick Man was Native American.

Jelderks had seen the legal notice in the paper and was familiar with Barran's position that NAGPRA did not apply to Kennewick Man. "Do you take it that that announcement and that decision, Miss Barran, does stand for a conclusion, on behalf of the Corps of Engineers, that the Act does apply?"

Barran referred to Lieutenant Colonel Donald Curtis's affidavit and read from it: "'I determined this individual to be subject to the inadvertent discovery provisions of the Native American Graves Protection and Repatriation Act.'"

"But," said Jelderks, "do you feel that is a binding determination by the defendants?"

"We do." Barran pointed to the corps' legal notice of intent to repatriate as further evidence. It said, "We have reached the conclusion that these are the remains of individuals of Native American ancestry."

Jelderks had carefully read NAGPRA in search of evidence that Congress intended the act to apply to ancient remains. "I couldn't find anything else, in anything I read, that would indicate that Congress or the drafter of the regulations really contemplated a 9,000-year-old skeleton."

Barran agreed. When it passed NAGPRA, Congress had in mind Indian skeletons that were on museum shelves and were clearly linked to particular tribes. The aim of the law was not to wall off scientists from identifying remains that date back thousands of years, much less to preclude them from study. "This is one of those things that is so unusual that we end up before you today," Barran said, "because the application of the law seems to do real violence to what we are faced with."

Jelderks focused on the word "indigenous." It appeared throughout NAGPRA and in the regulations explaining how to enforce NAGPRA. Before convening the hearing he had looked the word up in his dictionary, which defined "indigenous" as "occurring or living naturally in an area, not introduced." The words "not introduced" stood out. The phrase suggested that anybody who migrated here, even if by land bridge thousands of years ago, was not truly indigenous. The term "indigenous," it seemed to him, could be accurate only as applied to a specific date in history. But when Congress drafted NAGPRA it did not identify a point or a cutoff mark to define "indigenous."

He looked up from his notes. "Is there anything that is truly indigenous, in the scientific world—that we have today, if we go back to the beginning of time, or if we go back to the Ice Age?"

Barran hesitated. The question seemed far beyond the scope of a courtroom. "I don't know that I can answer that scientifically," she said. "But I think your observation is right, you have to pick a particular point in time to decide whether something is or was indigenous."

The issue intrigued Jelderks. "We can go back to when the first Europeans came to the shores of what is now the United States, back in Jamestown. Is that what we mean when we are saying indigenous?"

As Schneider watched, he tried to figure out where Jelderks was headed with this line of questioning. He could see the Justice Department arguing that the Europeans who migrated to Jamestown were not indigenous because they did not originate here, and there were other people here before them—Native Americans. But thanks to Kennewick Man, the same argument might apply to tribes that migrated to the Americas via the land bridge. They had not originated here, and Kennewick Man seemed to suggest that he had been here nearly ten thousand years ago.

Jelderks had a reputation for being very scholarly, very inquisitive. Barran sensed he was fascinated with the philosophical and scientific issues raised by this case. To her, that was a good thing.

"Your Honor, I think they picked a word and a concept and didn't realize, because they didn't have Kennewick Man in mind when they wrote this statute. It makes sense if you think about 1,000 years ago or 2,000 years. But to talk about something being indigenous when you are dealing with 9,000 years ago, and a major geological upheaval [taking place] in this area just doesn't make sense."

"Let's talk about the Constitution for a moment," Jelderks said. "I am having a hard time finding a constitutional right in your clients."

Barran suggested that under the First Amendment, the scientists had a right to know Kennewick Man's true identity. She also argued that their due process rights had been denied. She knew the First Amendment did not really protect a person's right to obtain knowledge. But she was sure her clients had been denied any access to the process of appealing for access to Kennewick Man.

"We usually talk about due process rights in relation to property rights," Jelderks said. "You are suggesting a due process right in relations to speech-slash-information?"

She was unsure how to classify a right to knowledge under the First Amendment. But Barran knew that due process is a limit on the actions of the government. In this case, the scientists were convinced that Kennewick Man was not a Native American. Yet the government was denying them the chance to prove it, opting instead to repatriate without any evidence of Kennewick Man's ancestry. "They would feel a lot more comfortable if the Army Corps had said not just 'we have to make up our mind,' but, rather, 'we will permit you a process to make a challenge' and 'we will give you access to information.' They have not done that."

Jelderks turned to Tim Simmons and Daria Zane, U.S. attorneys representing the Army Corps of Engineers. Their aim was to dissuade Jelderks from issuing an injunction that tied up Kennewick Man. Jelderks asked them to explain why they felt this case did not require the court to step in and block the transfer of the remains.

Zane spoke first. "On behalf of the United States and federal defendants we do not believe that there is any immediate irreparable injury that the plaintiffs could suffer."

Jelderks immediately interrupted. He wanted to know if she

thought NAGPRA applied to Kennewick Man. "It seems that the defendants and the Corps just presumed that the Act applied without there ever being any findings made to support that," he said. "If the Act does not apply, what rules apply to the disposition of the skeleton?"

Zane insisted that even if NAGPRA did not apply in this case, Kennewick Man still belonged in the hands of the Army Corps because he had surfaced on federal land. Under ARPA, Zane argued, Kennewick Man was the property of the United States and subject to the authority of the Army Corps of Engineers. "So either way we go," she said, "they [the remains] are with the Corps of Engineers."

Zane could not resist weighing in on Jelderks's earlier question to Barran about the word "indigenous." "On the issue of the definition of *Native American*, I think we have to look at the definition under NAGPRA," she said. "And NAGPRA says *indigenous to the United States*. Indigenous means of or connected with. It doesn't limit it to present day Native American tribes.

"If you look at it in the context of these remains, you have not only that they were located in an area that was known to be . . . occupied by Native American tribes, but you have that they are 9,000 years old. Once it was raised that these were 9,000 years old . . . and when the evidence became apparent that this was not an 18th-century settler, it was then that the Corps said, 'NAGPRA applies. We must take possession.'"

Jelderks stopped her. "But I don't know if it's quite as simple as saying this skeleton is 9,000 years old; therefore, it's a Native American." The law required proof that Kennewick Man descended from a tribe or people indigenous to the United States. He suspected that if one went back far enough in history, at some point the land that we now call the United States was void of any human beings. And if the first humans arrived here from somewhere else, were they truly indigenous? To Jelderks the issue needed to be resolved. But that question was for another day. He was more immediately concerned with Barran's insistence that the government had denied the scientists any procedure to appeal to the Army Corps for access to the skeleton.

Zane denied that claim. She said all the scientists had to do was file a claim to Kennewick Man under NAGPRA.

"But," Jelderks interjected, "if they filed a claim under the Act, they are conceding, in effect, that the Act applies." Barran and Schneider's argument was that NAGPRA did not apply. And even if they did file a claim under NAGPRA, it would be immediately dismissed. Only Native Americans are entitled to recover remains under NAGPRA.

Jelderks had one more question for Zane. "Since your argument appears to be that there is, in effect, a conclusive presumption that this is a skeleton of a Native American, based on the age, at what point in time does that conclusive presumption attach? And would we have the same result if it was a 100,000-year-old skeleton as a 9,000-year-old skeleton?"

The Justice Department's view of NAGPRA held that it was safe to assume that any old skeletons found in America were Native American. But at what point did the presumption kick in? Anything older than two hundred years? One thousand years? Three thousand years? And was it safe to assume that something one hundred thousand years old must be Native American?

Zane did not have an answer. But she assured Jelderks that Kennewick Man would not be repatriated until the corps determined what to do with him. She offered no specifics on what the decision-making process would entail.

The time had come for Jelderks to decide whether to issue an injunction. As a general rule, judges are reluctant to intervene in federal government agency action until a decision has been finalized. It now appeared that the Justice Department was saying that the Army Corps' decision to repatriate was being reconsidered. This removed any immediacy for the judge to step in.

"Ms. Zane, if you will assure me that . . . 14-day notice will go to Ms. Barran, I see no reason for the court to get involved in it at this period in time."

"Yes, absolutely, Your Honor."

Barran rose to her feet. A verbal promise from the government was not good enough. She wanted it in writing. Barran asked Jelderks to issue an order mandating that notice of any future decision to repatriate Kennewick Man be given fourteen days in advance.

"Based upon the government's representations, I will order that they comply with their agreement to give you 14 days' notice prior to any transfer of the remains in question."

Barran had one more request. She wanted to make clear that her clients purposely had not filed a claim to Kennewick Man under NAGPRA because they believed the legal reach of NAGPRA did not extend to Kennewick Man.

"Your claim simply is that the Act does not apply, and you want the record to be clear, with the government, that you have formally stated that position in a timely fashion."

"Yes, sir," Barran said.

Jelderks turned to Zane, who acknowledged Barran's argument but repeated what she had said earlier. Even if NAGPRA did not apply, the skeleton was still the property of the United States. "And," she said, "there is some concern that the United States, at this point, does not have all those remains returned to them that were in the hands of some of the plaintiffs. And we would like to have the court order that those remains be returned to the United States, the Corps as custodian."

Stunned, Barran and Schneider looked at each other. None of the scientists they represented had ever laid eyes on Kennewick Man, much less hands. They had not been granted any access to the remains. Schneider suspected something was amiss. Why else would the government, out of the blue, raise questions about missing bones?

"You have noticed I haven't asked any specifics about the skeleton," Jelderks said. "I have no idea whether something like this is of value in the world in general, whether it might be subject to thievery, or something happening to it."

Peeved at Zane's insinuation that the scientists might have stolen bones from the skeleton, Schneider reached for the microphone. "Your Honor," Schneider interjected, "can I address that first point about the remains that are in the hands of a laboratory?"

"Yes."

"It appears the Army Corps is not providing any better information to its counsel than it is to us," quipped Schneider.

"Do you have adequate information to represent to me that your clients are not in possession of any skeletal remains?"

"Yes, Your Honor," said Schneider. "I can categorically assure you that they are not in possession of any of the remains relating to this individual."

"I think that takes care of it then," said Jelderks. "Thank you for coming in on short notice, all of you. At some point in time we will be having a status conference. Anybody have any questions before we close?"

No one spoke.

"Okay. That's all for today then."

Barran looked at her watch: 3:10. She and Schneider had not convinced Jelderks to issue the injunction halting Kennewick Man's transfer. But the end result was essentially the same. Kennewick Man was not going to the tribes, at least not yet. And the government was under court order to give the scientists fourteen days' notice before making any repatriation. Barran was pleased. The government was now warned that a federal judge had his eye on them, and that any action from here on out would be brought to the court's attention and subject to examination at another hearing.

Kennewick Man is safe, she thought.

24

STAND AND FIGHT

October 24

Alan Schneider could not have been more pleased at how the previous day's hearing went. He sensed that the judge would keep close tabs on Kennewick Man for the time being. Schneider was more immediately concerned with losing his clients. The memo that Owsley and Stanford had received from Lauryn Grant on stationery from the Smithsonian's legal department said, "Our office has advised you to withdraw as plaintiffs in the above referenced litigation."

Although the memo didn't expressly state it, Owsley and Stanford felt they were being threatened with termination. But Schneider had a hunch that the Smithsonian had no such intentions. The Justice Department, he figured, knew it had inherited a terrible set of facts and was trying to avoid at all costs having a court examine the legitimacy of the Army Corps' decision to repatriate Kennewick Man. The mere presence of Owsley and Stanford in court undermined the credibility of the corps' decision. They had world-renowned expertise in prehistoric remains and were associated with the Smithsonian, the government's institution for collecting, storing, and studying prehistoric artifacts and remains. Yet the Army Corps had completely avoided Owsley and Stanford when it deemed Kennewick Man a Native American worthy of repatriation. In fact, the corps had actively worked to block Owsley and Stanford from seeing the skeleton. The scientists' willingness to speak out about the scientific errors being perpetrated by a fellow government agency put the Justice Department in a precarious position—it had to argue against the government's best scientists.

Lacking a legal basis for requiring Owsley and Stanford to withdraw from the suit, the Justice Department had resorted to political pressure.

Schneider called Owsley to discuss what to do next.

"Doug, my impression is that the Smithsonian is not totally behind this attempt to bump you from the lawsuit," Schneider said. "They are relaying what the Justice Department is saying. We have to keep nursing this thing along."

To Schneider, the longer Owsley and Stanford could fend off the Smithsonian from officially removing them from the case, the more time the institution would have to realize that its public credibility as a science institution would be undermined by siding with the Justice Department.

Owsley wanted to keep stalling. But he and Stanford were unsure what to do about Lauryn Grant's memo. She had spelled out the institution's wishes that both scientists withdraw at once.

"Let's respond to her memo with another memo," Schneider said. "We can put this back to them with some questions. You're entitled to get clarification on some of the things she said, as well as ask some questions. Let's do it in writing and get them to respond. Meanwhile, the days keep ticking off."

Over the weekend, Schneider drafted a confidential letter to Owsley, suggesting to him what to put in a memorandum to Smithsonian lawyer Lauryn Grant. After reviewing Schneider's letter, Owsley wrote his memorandum. "Neither Dennis [n]or I are attempting to cause difficulties for the Smithsonian Institution," he wrote to Grant, before launching into the science-based motivations behind their decision to sue. "I have carefully reviewed your letter and am uncertain about some issues that are mentioned."

He listed those issues for Grant.

"The memorandum refers to 18 U.S.C. 205 which is said to bar federal employees from acting as an agent or attorney in any claim or 'covered matter' against the United States," Owsley wrote. "As I am not a lawyer, I don't even know where to look up that particular statute, but I can assure you that Dennis and I don't see ourselves as agents or attorneys."

Owsley also listed some questions.

"By forcing us to withdraw as private citizens," he concluded, "it makes it look like there is some sort of government conspiracy to hide the truth. Your help and advice [are] greatly appreciated."

After having Schneider review his memo, Owsley initialed it and dated it October 26. To buy himself more time, rather than walking the memo over to her office, he put it in the Smithsonian's interoffice mail system, which usually moved slower than the U.S. Postal Service.

Barran and Schneider were concerned about the pressure being applied to Owsley and Stanford. Owsley, in particular, was the engine driving the lawsuit. Without him the whole case could evaporate.

"What are we going to do about this?" Barran asked Schneider.

Angry, Schneider suggested a suit against the government for interfering with the plaintiffs.

"There are legal issues that Doug and Dennis can raise if they want to," said Barran, calling on her background in labor law. Barran could go to a federal judge and say that the federal government was interfering with and harassing the plaintiffs, and interfering with something that is now a federal court lawsuit.

But Barran felt that decision was a personal choice that only Owsley and Stanford could make. Filing a harassment complaint might only make matters worse in the long run.

Schneider agreed. On November 4, Schneider and Barran sent Justice Department lawyer Daria Zane a letter.

"We have been informed that an individual from the Environmental Section of the Department of Justice has contacted the Smithsonian Institution to express objections to Drs. Owsley and Stanford's participation as co-plaintiffs in this action," they wrote. "We are very concerned about this matter and we wish to make sure there is no misunderstanding. Accordingly, please confirm or deny whether anyone from the Environmental Section, or to your knowledge any other section of the Department of Justice, has contacted the Smithsonian."

The next day Barran received a voice-mail message from Zane:

"Hi, Paula Barran. This is Daria Zane. I have talked to the attorney Steve Carroll in our office. And in no form or fashion has the Department of Justice directed, asked, or in any other fashion asked the plaintiffs to remove their name from the complaint. I do have the name of someone at the Smithsonian that you can call and talk to about this matter. It's Lauryn Grant at the Office of General Counsel."

Zane's message directly contradicted the memo sent to Owsley and Stanford. The Justice Department was saying one thing and doing another.

"This is smoke," Barran said to herself. She instructed her secretary to transcribe Zane's message.

November 7
Smithsonian Institution

Should they stay or should they go? Ultimately, the decision whether to allow Owsley and Stanford to go forward in the lawsuit had landed on the desk of museum director Robert Fri. A seasoned, high-level administrator both in the private sector and in several federal agencies, Fri had only a few minutes until Owsley and Stanford and lawyers from the Smithsonian's legal department assembled in his office to discuss the matter. He had a couple of questions on his mind as he awaited their arrival.

He knew federal employees typically couldn't sue the government, but exceptions existed. He didn't know whether this case qualified as an exception.

Nor did he know whether Congress had intended for NAGPRA to apply to remains as old as Kennewick Man. The law seemed silent on that question. The fact that Congress had neglected to address the issue of ancient remains in NAGPRA didn't surprise Fri. He had worked under Presidents Nixon and Ford as a senior administrator in the Environmental Protection Agency. During his tenure at the EPA, the agency routinely filed lawsuits and had lawsuits filed against it in an attempt to resolve problems that didn't get resolved cleanly in a federal law. Fri felt that Congress had dodged the thorny issue raised by ancient remains that have scientific value on the one hand, but are also regarded by certain Native American communities as having high cultural value.

As his door pushed open, he didn't know what his decision would be. Attorney Lauryn Grant and her boss, John Huerta, lead counsel for the Smithsonian, filed in. Stanford and Owsley soon followed. After everyone sat down, Fri asked Owsley to explain why he felt litigation was necessary.

As Owsley talked, Stanford nodded his head in agreement.

It seemed to Fri that it was up to the Smithsonian and consistent with its mission to find a way to advance the scientific side of the argument—knowing that others would advance the other side of the argument—and see what happens.

Fri looked at Grant and Huerta. He had one question. Was there a problem of policy or law that got in the way of Owsley and Stanford being plaintiffs in this case?

Grant reiterated the Justice Department's position as outlined in her previous memo to Owsley and Stanford.

"Look," said Stanford. "I talked to our lawyer and he read the memo. He found that the Justice Department's position isn't valid. His opinion on that issue is that the Justice Department is flat wrong and that we don't have to get out of the lawsuit."

Fri had already decided that if a statute prohibited Owsley and Stanford from suing, he would ask them to withdraw. But he sensed no such prohibition existed. He turned to Huerta. "Is this a legal issue?"

Huerta paused, then conceded that he didn't think they could really ask Dennis and Doug not to be involved. But he would prefer that they weren't.

His arms folded across his chest, Stanford looked at Fri, wondering if he would back science or take the politically safe position and ask them to withdraw.

Fri looked around the room. "If there is no legal prohibition," he said, "then I'm of the view that as a matter of policy, this intervention by two Smithsonian scientists is appropriate and the right thing to do."

Owsley could not conceal his glee. He looked at Stanford and raised his eyebrows. Pleased and relieved, Stanford remained straight-faced and silent.

"Our role in life here at the Smithsonian is scientific research," Fri continued diplomatically. He believed that if they were going to get this dispute over ancient remains sorted out, it required legal action. "Certainly we respect the traditions and cultures and beliefs of Native Americans," he said. "But here we are confronted with an issue.

"There are two sides to it. It's not clearly resolved in the legislation. We're going to have to resolve it through litigation. I think they should go forward."

A relieved Owsley and Stanford left Fri's office. Their jobs were safe.

25

INTENT

February 3, 1997
U.S. District Court
Portland, Oregon

Their shoulders nearly touching, Barran and Schneider sat at a table. A microphone was positioned in front of them, amid stacks of legal files and notepads. Three feet to their right, lawyers from the Justice Department sat at an identical table. Both parties anxiously anticipated Judge Jelderks's entrance into the courtroom. Three months had passed since the initial court hearing, which halted the transfer of Kennewick Man to the tribes. Without coming forward with its plans for the skeleton, the Justice Department had filed a motion asking Jelderks to dismiss the scientists' case. Barran and Schneider came prepared to challenge the motion.

Anxious, Barran glanced to her right. She found it ironic that representatives from the Army Corps of Engineers were huddled around the Justice Department's table. The corps is an agency that is supposed to issue unbiased decisions. Yet Justice Department lawyers who were sworn to contest Barran and Schneider were advising them. It hardly appeared unbiased.

When Jelderks entered the courtroom, he greeted the lawyers, then asked the Justice Department to make an opening statement.

Attorney Daria Zane stood up. "First of all," Zane began, "I think it's important to understand exactly what is the issue before the Corps of Engineers and how NAGPRA operates on that issue. NAGPRA involves the disposition of human remains to groups. And these groups are Indian tribes or native Hawaiian groups that have a relationship with those remains; they are culturally affiliated.

"Where there is a remain that's discovered, and a tribe comes in and makes a claim and says this is related to us, it is disposed of to that tribe. The custody is transferred.

"And the procedure that NAGPRA does this is if a remain is found on federal property, it imposes some responsibilities on the federal agency. And one of those responsibilities is to notify tribes that are likely to be culturally affiliated. The purpose of this is so that the tribes can go and determine if they believe they are culturally affiliated to make a claim."

As Zane spoke, Schneider scribbled on his legal pad, noting what he thought were errors in her understanding of NAGPRA. Barran sat still, listening to Zane explain how after Kennewick Man was discovered, the Army Corps notified numerous Indian tribes that occupy land closest to the discovery site. One tribe, the Umatilla, claimed that the site of the discovery was on land that had been ceded to them by the United States in a treaty signed in 1855.

"Based on that information," Zane said, "the Corps found that it was likely that those tribes should be the legal custodian."

The corps then published a notice of its intention to repatriate Kennewick Man to the tribes. "And what happened," Zane said, "is in response to that notice of intent, there were a number of other claims filed. In addition, there was information that was derived that indicated that maybe these weren't ICC [Indian Claims Commission] judgment lands, and we need to go back and look at this. And so they [the Corps] have had this proposed action, they got comments, they got other claims, and they now are in the process of going back and looking at that."

Barran glanced at her typed notes, listening for whether Zane said anything that wasn't covered in her planned response. So far Zane had said nothing unexpected. She told the court that whether the lawsuit had been filed or not, the same process would have been followed, and the remains would not have been turned over to the tribes prematurely. "Now the plaintiffs have said that they have been precluded since they can't claim under NAGPRA," Zane said. "But the point is not that they can't claim. The Corps currently has before it a decision as to whether or not these should be transferred under NAGPRA, [whether] there is cultural affiliation. And if there is, the remains should be transferred to the tribe with the closest [cultural affiliation]. And the point is moot.

"But if the remains are not [culturally affiliated], they are not going to be disposed of under NAGPRA. And there is nothing to prevent a group from coming in and establishing that they have a link under some other statute.

"But in any event, the fact of the matter is this is all premature because there has been no decision by the Corps at this point. There has been no decision by the Corps to transfer custody under NAGPRA. And until that time, this isn't fit for review."

Judge Jelderks looked toward Barran and Schneider. "Miss Barran," he said.

Barran stood up and adjusted the microphone, pointing it upward.

"Good morning, Your Honor. Listening to the government's opening argument, you would assume that once a claim is made, the remains would automatically be transferred."

"Miss Barran," the clerk interrupted, struggling to hear her, "why don't you be seated and pull the microphone to your face."

Perplexed, Barran stopped. In the countless oral arguments she had made throughout her career, she had always stood—it was protocol when addressing a judge. "I am not sure I can possibly do oral argument seated," she said, sitting down. "Old habits die hard."

Placing a Federal Express box under the microphone in order to elevate it, she pulled the microphone to her. "Right now the attorneys in the case are taking a different position from what their clients have been taking," continued Barran, referring to Zane's opening argument. "The position is: 'we are under NAGPRA'; 'we are attempting to find out cultural affiliation—it is only one decision, but we are operating under it'; and 'we have invited other claimants to come forward.' Well, if they are inviting our clients to come forward, they are violating the very law that they are proceeding under, because we are not allowed to come forward under NAGPRA.

"The first thing that we believe that we can prove," she said, "is that there was a predetermination made of Native American status. You don't get into NAGPRA until you have first decided that you have something before you that is Native American. The government has been behaving as if this is clearly a Native American skeleton. That is why they are off onto cultural affiliation already."

Barran and Schneider had brought with them to court a series of documents to prove their point. "The first is the notice that was issued," Barran said, reaching for a photocopy of the intent-to-repatriate notice that the corps published in the newspaper after seizing Kennewick Man from Benton County officials. Barran had highlighted a passage in yellow. "It says—and I'm quoting—'officials of Walla Walla District have determined the human remains listed above represent the remains . . . of Native American ancestry.'" Barran looked up at Jelderks. "We challenge that determination, but that has already been determined. That is why we are under NAGPRA, according to the government."

She looked back down at the notice, locating another highlighted passage. "Further language in the notice says, quote—'officials of the Walla Walla District have also determined that there is a relationship of shared group identity which can be reasonably traced between the human remains and five Columbia River Basin tribes and bands'—end quote. There is no reference to *may be* or *possibly*. They have determined."

Barran flipped to another document, a court transcript. "There were also statements made in open court in October," she continued, referring to the original hearing. "Counsel for the government said there, quote—'there clearly was a basis for saying these are Native American.'

"Our plaintiffs challenge that determination, because we do not believe that there was a basis for saying the remains are Native American. Let me give you an example of something we just learned."

Schneider handed Barran a copy of a letter recently written by

Lieutenant Colonel Curtis from the Army Corps. "Colonel Curtis admits now that something that was said in the notice was wrong," Barran said. "The notice says that the remains were found on aboriginal lands. The letter admits that that apparently was in error. That seriously undercuts the basis for issuing the notice in the first place."

Barran put the letter down. "We believe that we can prove, on this record and under the complaint, first, that there has already been a determination that the remains are Native American, and, therefore, NAGPRA applies. And if the government is *not* there, then I question why we are under NAGPRA at all, because that is an essential predicate to the statute."

Barran reached for one more document. Looking at Jelderks, she continued. "You have, in the record from the injunction proceedings in October, an affidavit signed by Colonel Curtis. And he says in that affidavit—and again, I am quoting—'I determined this individual to be subject to the inadvertent discovery provisions of the Native American Graves Protection and Repatriation Act and implementing regulations.'"

Barran set the affidavit aside. "I *determined*. Not, 'It seemed to me possible.' Not, 'It seemed to me likely.' Not, 'It appeared.' But, 'I determined.' And then counsel, at oral argument in October, said the same thing. It was then that the Corps said NAGPRA applies.

"Our clients have been excluded from the process. The notice says 'any other *Native American tribe* which believes itself to be culturally affiliated with the human remains should contact . . .' That is a very far cry from what the Department of Justice has been saying, which suggests that we may come forward and make our own claim. Once we are under NAGPRA, the Corps is not allowed to consider our claim. We have no right."

As Barran concluded her argument, Schneider was itching to speak. In addition to teaching the NAGPRA law and its application to federal government employees, Schneider had studied the law more than anyone else in the courtroom. The law had become his hobby.

"Your Honor, if I may be permitted [to speak]. What's happening here is the government keeps making up the rules as they go along. And somehow, we always tend to lose when they make up the rules. The government is ignoring the fact that the notice of intent to repatriate does not *start* the process. It's, in effect, *the culmination* of the process."

After listening to all the arguments, Jelderks thanked the attorneys and promised to issue a ruling soon on the government's request to have the scientists' case dismissed. The Justice Department's argument was simple: The Army Corps of Engineers had the authority to decide what to do with Kennewick Man. The scientists' lawsuit was premature because the corps had not yet made up its mind regarding Kennewick Man. And a federal court could not intervene until after an action was taken.

Barran and Schneider had done all they could to convince the judge that the corps had made up its mind months ago, and had only held back because of the lawsuit. If the judge dismissed the case now, they feared, Kennewick Man would go back in the ground in a heartbeat.

26

SCIENCE EVOLVES

February 10, 1997
Jamestown, Virginia

In 1670 the Virginia Colony enacted legislation that made all African servants slaves for life. Over the following forty years, approximately five thousand Africans were shipped to Virginia by the Royal African Company. But the historical record is slim when it comes to the experience of Africans who entered the colony prior to 1670. Ship logs and correspondence between Jamestown colonists establish that the first Africans—twenty in number—arrived at the colony aboard a Dutch vessel in 1619. Historians believe that the Dutch obtained the Africans after plundering a Portuguese slave ship that had originated in Angola. The Portuguese had been trafficking slaves from Africa to the West Indies since the late 1500s.

But varying views exist over the status of the Africans who came to the colony after 1619 and before passage of the slave law in 1670. During that period the emergence of tobacco production required a massive increase in labor. An estimated seventy-five thousand whites came from the British Isles to the Virginia and Maryland colonies. Indications are that upwards of three quarters of them were indentured servants who were eligible for freedom after four to seven years of service in the tobacco industry. Far fewer Africans—some scholars suggest that five percent or less of the population was black—entered the colonies during this period. Census and muster records indicate that some Africans came to Jamestown with white households that had migrated from England. But historians disagree over the legal status of Africans at Jamestown prior to 1670, and whether the social conditions were distinguishable between servants and slaves. Archaeology has done little to inform the debate. The Parks Service excavated a couple of pairs of shackles in the 1930s. Alone, the shackles do not confirm blacks' presence. Shackles were used for prisoners of all races in the colony.

Owsley was not part of the debate. But the African skeleton he had discovered in the Jamestown Museum promised to shed light on the medical and social conditions of the time. Historians and archaeologists alike awaited his analysis. The moment had arrived. With his assistant Kari Sandness at his side, Owsley began. "Present are the remains of an adult male. Leave the age blank. Sex: male. Ancestry: African. The skeleton is nearly complete."

Wearing a white lab coat and latex gloves, he picked up the skull, running his fingers over the rough surface. "That's a classic feature of cranial syphilis," Owsley said. After giving Sandness a complete description of the skull, he examined the skeleton's other bones for syphilis. He saw a pattern. From head to toe, the skeleton's bones exhibited extreme roughening and irregular contours along the exterior surfaces. The bones also indicated a long-term chronic condition of bone formation, destruction, and remodeling. All the symptoms pointed to syphilis, the sexually transmitted venereal disease characterized by hard red lesions, ulcerous skin eruptions, and systemic infection that results in partial paralysis and brain damage that can lead to insanity.

As Sandness recorded notes on her laptop, Owsley picked up the skull again, noticing a foreign substance on the left side of the forehead. "Putty," he said, touching the rubberlike patch. "Somebody did a pretty good job with the restoration here." The putty, nearly identical in color to the skull, covered a key-shaped hole in the forehead. Owsley examined the hole more carefully. "This is a perfect little hole," he said. "And there are radiating fractures coming out from it. See that?"

Sandness put her hand on the skull, looking at the area Owsley pointed to.

"These archaeologists," he said to her, "when they dug, weren't expecting to hit a skeleton. Then they hit the skeleton. And they thought they damaged the skull, so they very neatly patched it up."

"What do you think caused it?" she asked.

"Actually, it looks to me like a gunshot wound."

Sandness said nothing.

Owsley summoned David Riggs. "Dave, we need to borrow this skeleton."

He showed Riggs cracks traveling away from the possible gunshot hole in the skull. "The way to prove it is if we can find the lead fragments embedded in the skull. But it's too complicated to study here. We need X-ray equipment."

Intrigued, Riggs agreed.

Portland, Oregon

Dismayed, Alan Schneider tossed aside the latest memorandum that the Justice Department had filed with the court. Still attempting to block the scientists from studying Kennewick Man, the government's lawyers had come up with a new strategy. The Justice Department was arguing that Schneider's clients shouldn't be allowed to study Kennewick Man because they were biased and used unscientific methods.

Schneider knew he represented some of the most talented anthropologists and geologists in the country. He knew that the government knew it too, since Owsley and Stanford worked for the Smithsonian. But he had to show that the government's argument was disingenuous.

Schneider called Owsley. "Doug, the government is essentially arguing that your studies are not scientific and that you're biased. We have to refute that."

"What do you need me to do?"

"Let's start with the bias issue. Tell me all the federal agencies you have done work for."

Owsley rattled them off. The State Department had dispatched him to Croatia to help that country identify war dead. The FBI had sent him to Waco to identify disaster victims. The U.S. embassy had used him to help identify American journalists in Guatemala. The Defense Department had brought him in to identify military personnel killed in Desert Storm. Schneider interrupted Owsley with laughter.

"Wait a minute, Doug. All these agencies are using you to identify skeletons in all these top-priority federal cases?" The irony amused Schneider. The government was relying on Owsley and his method of analysis in all kinds of cases, and trusting his judgments. Yet, the government was arguing in the Kennewick case that Owsley should not be trusted because his methods were not scientific.

Owsley explained that the irony was even richer than Schneider realized. Since the passage of NAGPRA, other federal agencies, including the Army Corps of Engineers, had routinely enlisted Owsley to examine Indian remains to make sure they were repatriated to the correct tribes. No one had examined any more Native American remains since NAGPRA's passage. In some cases, without Owsley's intervention, federal agencies would have repatriated Indian remains to the wrong tribes.

"Jantz and I have gone in and showed where Native American skeletons that were identified as Sioux were in fact Pawnee," Owsley said.

"Really?" Schneider said.

Mistakes were easy to make, Owsley explained. There are certain obvious skeletal characteristics that distinguish Native Americans from other human populations. However, the skeletal differences among tribes are very subtle. But Owsley had personally studied more than five thousand Native American remains, more than any other scientist in the world. No textbook could substitute for his hands-on experience. His eyes were trained to spot small differences in human populations that were evident in the bones.

Schneider asked for more examples in which Owsley had prevented Native American remains from one tribe being mistakenly repatriated to another tribe.

Owsley instantly thought of a better example. He told him about the Jamestown burial of African descent that had been identified as Native American.

"How . . . how could someone misidentify a skeleton so that an

African could be mistaken for a Native American? Did the original scientist just blow it on the identification?"

Owsley explained. A professor named George Neuman identified this particular skeleton in the 1940s. Neuman was a very good physical anthropologist. And he used what were state-of-the-art techniques for his day.

"So how did he get it wrong?" Schneider asked.

"Today we just know so much more and the technology has advanced so much further along that we can make much more accurate identification."

The African skeletons from Jamestown highlighted the danger of arbitrarily repatriating old remains to tribes under NAGPRA without performing proper scientific techniques. "Little historical information is available for blacks prior to the eighteenth century," Owsley added. "These physical remains represent a century for which records are extremely rare, especially for American minorities."

Listening to Owsley, Schneider got an idea. The Jamestown case needed to be brought to Judge Jelderks's attention. It provided a powerful argument for why Kennewick Man had to be properly identified before being repatriated. "Doug, when can I get you to put what you just told me into an affidavit?"

"Well, I want to finish my analysis of the skeleton first, then produce a paper on the case."

By the end of February, Owsley had identified two more African skeletons, a man and a woman, in the Jamestown collection. He brought those two and the syphilis-ridden skeleton to the Smithsonian to have them X-rayed. The pictures of the skeleton with syphilis also revealed metal fragments embedded in the fractured bone. The pattern indicated that the bone had shattered on impact. The victim had been shot in the head at point-blank range.

Searching for motive, Owsley considered the colonial time period in which the death occurred. Brutality against a slave seemed a natural answer. But the bone pathologies didn't fit that scenario. It was the most advanced case of syphilis Owsley had ever seen. The subject had little hope of survival. Yet Owsley could tell from the bones that the subject had lived to approximately forty years of age, far beyond the life expectancy for someone whose bones demonstrated years of disease.

For a person to have survived that long with such an extreme expression of disease, Owsley reasoned, somebody must have gone through great lengths to keep him alive, and nursed him. He looked at the wound location again, off to the side of the forehead. A scenario started to emerge. When the caregivers knew that their health care had no prayer of saving him, they killed him so that he wouldn't have to suffer anymore. Euthanasia, Owsley figured, could very well have been the motive behind the shooting.

The burial style also suggested something other than murder. The body had not been hastily discarded, as was standard practice in the seventeenth century for disease-ridden individuals, wherein shallow pits were used for graves. The practice was sometimes literally to pitch the dead into the pit, causing corpses to land on their stomachs, the bodies halfway twisted, with arms and legs sprawling in random directions. Soil was then thrown on the bodies. Out of fear, no one wanted to touch the diseased bodies.

This African skeleton had received an entirely different burial. He had been perfectly laid to rest in a grave, hands properly clasped and tucked under his chin, the body in a semi-flex position. If he had been murdered or discarded, someone would not have taken the time to meticulously arrange and inter the body. Even in the end, it seemed that somebody had done the best that he could do for the African. "The gunshot wound to the forehead was likely precipitated by severe dementia caused by advanced syphilis," Owsley wrote in his report.

Between the well-preserved, syphilis-riddled skeleton and the two more fragmented African skeletons, Owsley confirmed an African presence at Jamestown prior to the slave law. But the skeleton with the gunshot wound particularly interested Jamestown archaeologist Bill Kelso. Its medical condition, manner of death, and burial suggested a life different from that of the typical slave. The possibility complemented an observation he had made with regard to Indians and Jamestown colonists. Since beginning his preservation and restoration project at the colony in 1993, Kelso had found evidence that Indians lived inside the fort with whites, as was suggested by the stone arrow points, as well as the stone chips generated in the manufacturing stage found there. Kelso was convinced that the colonists would not make stone arrow points. They didn't know how to make arrows. Nor did they have the need; they had access to metal. The prospect of Indians and colonists living together—even if only for a brief period before hostilities broke out—intrigued him. Owsley's observations bolstered the notion that Indians, blacks, and whites all lived at the fort. To Kelso, this suggested that America hasn't just evolved into a racially diverse country; it began as one.

When Owsley finished working on the Jamestown skeleton, he called Schneider, telling him he could start drafting an affidavit.

"Alan, the Jamestown case is an example of why we have collections and why we study collections," he told him. "The techniques that we use—the forensic experience—were not part of past scientists' repertoire. This is a perfect example of how science evolves."

With Schneider's help, Owsley completed his affidavit.

> I, Douglas Owsley, being duly sworn, do depose and state as follows. I give this affidavit to respond to defendants' attempt to create the impression that plaintiffs are not qualified to provide scientific data and input in the present case. Dr. Jantz and I have examined and documented more than 350 skeletal remains for NAGPRA purposes for . . . federal agencies and more than fifteen museums, universities and state agencies. In the course of these NAGPRA projects, we have been instrumental in correcting the record in a number of cases where skeletal remains were misidentified to geological age, provenance, race or tribal affiliation. One recent incident involved three burials at the Jamestown Colony that had been identified as Native American. In fact, they proved to be Africans. They represent the earliest known Africans in the British North American colonies. If we had not reidentified these individuals through visual observation and comparison of their cranial measurements to Dr. Jantz's reference data, they would have been subject to repatriation and African-Americans would have lost an important part of their heritage.

Schneider filed Owsley's affidavit with the court in Portland on May 23, 1997.

27

VIRTUAL REALITY

November 1997
Carson City, Nevada

Riding in the car as Smithsonian photographer Chip Clark drove from the airport in Reno to the Nevada State Museum, Owsley looked at a computer-generated map of the United States. It indicated the archaeological sites where the oldest skeletons in the country had been found. Owsley reviewed them on the map.

Buhl Woman, a 12,825-year-old teenage girl found in a rock quarry in Buhl, Idaho, in 1989. Almost no study was performed on her before she was turned over to the Shoshone Bannock tribes in 1991 and reburied. After the fact, Owsley and Jantz obtained photographs of the skull and the limited data collected before reburial. From the photographs they calculated the skull's true dimensions and determined Buhl Woman bore no affinity to modern American Indians.

Minnesota Woman, an 8,775-year-old adult woman discovered in Pelican Rapids, Minnesota, in 1931 during a highway construction project. Owsley and Jantz examined it, confirming that the skeleton shared no affinity to modern Native Americans. DNA testing was also performed. Regardless, the Sioux tribes of South Dakota reburied it on October 2, 1999.

Browns Valley Male, a 9,710-year-old male found on private land and donated to Hamline University in Minnesota. The Minnesota State

archaeologist decided to repatriate it to the Sioux tribes of South
Dakota for reburial. Owsley and Jantz studied its skull features. It
showed no affinity to modern Native Americans. The Sioux buried it
on October 2, 1999.

The Hour Glass Cave burial, an approximately 9,140-year-old male
in his forties who was discovered in the Colorado Rockies in 1988.
With almost no research conducted, he was turned over to the Southern Ute tribe and reburied.

Wizards Beach Man, a 10,825-year-old man in his thirties who was
discovered in Nevada around the same time the Spirit Cave mummy
was discovered. The skeleton remains curated at the Nevada State
Museum and has been claimed by Paiute tribes.

Grimes Point Burial Shelter. A female child less than ten years old
at the time of death. The skeleton dates back 10,825 years. She has
been claimed by the Northern Paiute tribes and is curated at the
Nevada State Museum.

Besides gathering data on these skeletons, Owsley and Jantz had
examined twenty-three other skeletons or partial skeletons between
8,775 years and 13,000 years old. Kennewick Man was not the oldest. But
he and Spirit Cave man were the most well-preserved intact skeletons of
the bunch.

✳ Kennewick Man had another distinction. He was the key to
whether the others got studied or buried. Owsley knew that if he and
his colleagues lost their suit and Kennewick was declared off-limits to
scientists and reburied, all the future discoveries of ancient skeletons
would soon follow. He had to win. He just had to.

Clark and Owsley pulled up to the museum. Inside, Amy Dansie
had the final report on the Spirit Cave mummy's diet. The contents of
its intestines had been thoroughly analyzed.

"So what was his last meal?" Owsley asked.

"Fish," she said.

Dansie gave him a copy of the published report on the mummy's
intestinal contents. It confirmed that the mummy had ingested fish, as
large amounts of fish bones, as well as some fish-eye lenses, some cal-

cium-carbonate nodules, and animal fibers were detected. The report also contained aerial photographs of water habitat around the discovery site where the Spirit Cave mummy died. "Tui chub are found in most water regimes in the Great Basin, as are suckers," the report said. "Based on this, a shallow water habitat may have existed around, or very near Spirit Cave."

The diet report further convinced Owsley that the Spirit Cave mummy's unique skull shape and size were not caused by radical changes in the environment. There has been radical change in human history in the last one hundred fifty years, both in diets and life expectancy rates. Such changes tend to translate into minor changes in skull size and configuration. But Owsley pointed out that what the Spirit Cave mummy was eating and what his parents were eating and what their parents were eating is probably not that different from what was found in his stomach. There might have been environmental changes that produced more fish or less fish. But nothing in the environment could account for the differences in skull shape between Native Americans and the Spirit Cave mummy.

Dansie had approved studies that went beyond determining the mummy's diet. In an attempt to better understand the mummy's facial appearance, Dansie had authorized a CAT scan of the mummy at a nearby hospital. Dansie had the CAT scan sent to a facility in Texas that uses a computer-driven laser—a stereolithograph—that reads CAT scan images. The laser beams reflect the image of the skull from the CAT scan onto a liquid plastic. The liquid plastic hardens as the laser hits it, ultimately forming a perfect reproduction of a resin skull.

Dansie gave a replica to Sharon Long, a reconstruction artist with a degree in anthropology. Long made a mold of the skull out of liquid rubber. From the mold, she produced a plaster cast.

Familiar with Long's work, Owsley contacted her to schedule a time to sit with her and develop the mummy's eye color, hair color, facial wrinkles, and skin pigment—details that would result in a virtual-reality bust of the Spirit Cave mummy.

Days later
Portland, Oregon

"What the hell's going on?"

Alan Schneider could not contain himself. He reread the Army Corps proposal, making sure it said what he thought it said. It did. The corps planned to truck tons of rock and soil to the Kennewick Man discovery site, then drop it on the site with the aid of cranes and helicopters. After covering the site, the corps planned to plant dozens of fast-growing trees on it.

"Cover the site?" Schneider said to himself. "What is this bullshit?"

For months, Schneider had been trying to persuade the Army Corps to let two of the plaintiffs—both nationally renowned geologists—take sediment samples from the soil beneath the discovery area. Soil and other organic material likely held clues that could corroborate the carbon date assigned to Kennewick Man, as well as insights into the skeleton's culture and environment. The geologists offered to cooperate with the tribes and to share with them the collection and analysis of soil samples.

One of the tribes expressed an interest in examining the soil. But the Umatilla had protested any geological tests on soil around the site. "Our preference is that nothing be done, that no digging be done, that the area be left alone," a tribal spokesman told the corps. Schneider found the request troubling. The site was not a tribal burial ground. It wasn't even tribal land. It was a public park. And the soil beneath the surface could hold important archaeological clues to Kennewick Man's identity.

The corps rejected the geologists' request for soil samples. Instead, the corps said it planned to bring its own geologists to sample the soil. The corps had sent Schneider a copy of their soil-testing protocol. It contained an oblique reference to burying the site after the corps received sediment samples.

Schneider immediately called the Kennewick area's congressman, Doc Hastings, and Washington's U.S. senator Slade Gorton for help. He explained to their staffs that the corps planned to cover up the site with rocks, soil, and trees, forming an impenetrable barrier against future archaeology or scientific research. Any beneath-the-ground clues to Kennewick Man's identity would be forever lost. Hastings and Gorton agreed to intervene and asked Schneider to draft a proposed bill to protect the site.

While Schneider drafted the bill, the corps brought its geologists there in December 1997. They discovered a layer of volcanic ash that dated back 6,800 years. The government's geologists concluded that the Kennewick Man bones were in soil that was beneath the ash, confirming he exceeded 6,800 years in age.

Schneider finished his draft legislation right after the government geologists finished their survey of the site. He sent the proposed bill to Washington as the corps made plans to cover the site. Legislative aides from Hastings's and Gorton's offices rewrote the bill and had it introduced in the House of Representatives and the Senate. By the spring of 1998, Schneider received word from Washington that the bill had passed both the House and the Senate. All that remained was for a Congressional committee to resolve some minor differences in unrelated provisions of the bill.

Schneider called Barran to tell her the news. The site had still not been covered, and the bill made it illegal for the corps to follow through with their plans.

While they realized that the bill had still not been signed into law, the legislative staffers on Capitol Hill assured Schneider that the bill had no chance of being vetoed.

"We beat the government to the punch," Schneider said. "We got the bills passed. They won't cover the site."

Barran said little before hanging up with Schneider. She had personal concerns that made it difficult to focus too much on the legislation. She had decided to leave her law firm and form her own and was in the process of finding office space and hiring attorneys. That was the least of her problems. Her husband, Richard, who had agreed to join her firm, had been diagnosed with colon cancer. The uncertainties of his future made the practice of law seem less significant. She counted on Schneider to carry the case until her new firm opened and Richard completed his biopsy treatments.

• • •

Schneider had barely got the site protected when he read the quarterly status report that the Justice Department had filed with the court. In it, the government disclosed that an expert it had hired to analyze the condition of Kennewick Man had reported that the skeleton's femurs were missing. "How on earth can that be?" Schneider wondered. "Femurs are not tiny bones."

He read on. The government told the court in its memo that it had examined the photographs taken by Jim Chatters on the night he boxed up Kennewick Man in his basement. Those photographs confirm that the femur bones were present at that time. The government then examined the inventory sheet that Umatilla archaeologist Julie Longenecker filled out the day she and a host of tribal members were present after Kennewick Man arrived at the Battelle storage facility. The inventory did not mention the femurs. A five-day gap existed between the day Kennewick Man arrived at Battelle and the day Longenecker took the bone inventory. "Several of the femur fragments noted in Dr. Chatters' photographs are not with the remains," the memo read. "Defendants are in the process of seeking additional information in order to determine the disposition of the femur fragments, including contacting representatives from the Benton County Coroner's Office and counsel for Dr. Chatters."

Schneider put down the memo. His mind raced back to the first court hearing, when a Justice Department lawyer implied that the scientists might have taken some bones. *When* were the femurs taken? he wondered, reaching for the case file on the Kennewick Man litigation. He flipped rapidly through the record, looking for the date on which the judge had ordered the federal government to properly preserve the scientific integrity of the skeleton. It was June 27, 1997. He reread the judge's order. It expressly instructed the government to take necessary steps to ensure that the highest standards of curation would be employed throughout the course of the litigation.

Schneider put the order down. If the femurs were taken after June 27, 1997, then the government had violated a court order. If they were taken before the court order, then why had the Army Corps only now discovered them missing?

Schneider sensed foul play. Besides the skull, the femurs are the ✳ most crucial part of a skeleton, containing more evidence than any other bone in the human anatomy.

Schneider called Barran.

"Part of the skeleton has been heisted," he said.

"*What?*"

"The femurs have disappeared. According to the government, they're gone."

"Where does this take us? What are our issues?"

"Well," Schneider said, "if we don't do something here, we may not have any skeleton left to measure if we win the lawsuit."

Together they considered who had both motive and opportunity to take the femurs.

"One of the first things the government is going to do is point to Chatters," Schneider said. "But this makes no sense. First of all, holding back the femurs would reduce the scientific value of Kennewick Man. And Jim has a strong professional and personal interest in preserving the importance of the skeleton."

Barran agreed.

"Second," Schneider continued, "he could never do anything with the bones. Once he was obligated to turn them over to the government, he couldn't pop up and say, 'By the way, I've got these other fragments.'"

Barran laughed.

"Then there's the coroner," Schneider said. "He has no motive to keep them. He could lose his job over this. What does he have to gain by keeping them?"

Barran and Schneider figured that there were two likely occasions for the bones to disappear: during the skeleton's transfer from the coroner's office to the Battelle facility or during the five-day period between being deposited at Battelle and being inventoried at Battelle.

Schneider agreed to handle the investigation into leads on the missing bones. Barran had to focus on her husband, who was preparing to undergo surgery to remove the tumor in his colon.

28

THE COVER-UP

Its blades cutting through the dry spring air, a helicopter with a long steel cable bearing a bucket packed with rocks hovered over the shoreline where Kennewick Man had surfaced nearly two years earlier. Wearing a yellow hard hat, a construction worker standing on the shore guided the bucket until it dumped its contents on the shoreline, generating dust and drowning out the sound of a drum being pounded by an Indian who watched as the workers relentlessly dumped buckets of rock and fill atop the Kennewick Man discovery site.

"I feel good," the drummer said. "This is a good thing."

"This is preservation of our culture," said Umatilla Indian leader Armand Minthorn, on hand to watch the site covering.

The Earth Construction Company, hired by the Army Corps of Engineers, brought in tons of fill, enough to cover 250 square feet of shoreline. Russian olive trees—fast-growing trees that quickly establish deep root systems—were planted on top of the fill, further preventing any access to the geological time capsule below.

Hours later
Portland, Oregon

Schneider laughed when the press called and told him that helicopters were unloading landfill over the discovery site. But he didn't think it was funny. The corps had floated the idea of doing this earlier in the case, and members of Congress had disapproved. When he hung up, he pulled a letter from his file. Senator Slade Gorton and U.S. representative Doc Hastings had written it to Interior Secretary Bruce Babbitt. "We urge you to show greater respect for the legislative branch than has thus far been exhibited by the Corps of Engineers and put a stop to this needless destruction," wrote Senator Gorton and Representative Hastings, who also stressed the financial cost of burying the site. "We can see no reason to intentionally damage the site by depositing tons of rock and soil."

Schneider looked at the date: April 3, 1998, just three days earlier. The corps had acted before Congress passed the bill to protect the site. Schneider laughed again, grimly this time, then dialed Barran's phone number.

"Paula, they covered the site," Schneider said.

"Oh shit, really."

"The rocks are actually falling on the site. Can you believe that these people are actually defying Congress?"

"We'll never know what was under there," she said softly. "If you think of investigating K-Man as a big jigsaw puzzle, we now have a couple of holes. There are pieces that will forever be missing."

"This is outright *defiance*," Schneider said, raising his voice.

Schneider wanted to investigate. Barran encouraged him. She had come to believe that Schneider was as tenacious at pursuing legal facts as Owsley was at pursuing scientific ones. If there was anyone who could find out what prompted the corps's actions, it was Alan.

He did not disappoint. Relying on a paper trail of internal memos from the corps and other government correspondence that he tracked down, Schneider discovered that in early November an unidentified White House official ordered Lieutenant Colonel Donald Curtis to proceed with an "armoring project." The project's aim was to cover the discovery site before January 1, 1998. The Army Corps was given a budget of $200,000 to complete the task. When the legislation to protect the site passed the House and Senate, the corps told Congress it would comply with the legislation. But while the Congress enjoyed a brief April recess, the corps hastily ordered the armoring project to go forward. The decision was made at the highest level of the corps. The top brass knew full well that the move would generate a public outcry. But General Joe Ballard, the Commander of the corps, predicted that "the din will die out very quickly." The corps spent nearly $170,000 of taxpayer money to carry out the site cover-up.

When Schneider shared his findings with Barran, she was incredulous.

"I'm flatly appalled that the corps could walk in and do that knowing that there's a law that's almost passed to tell them not to," Barran said. "This is uncontrolled government."

When Barran hung up with Schneider, she wondered how the site cover-up would impact the litigation. She likened the situation to setting fire to the library at Alexandria. She knew that her clients were going to be gravely disappointed. But in terms of the litigation, she saw a potential upside to the government's tampering with the site; however, she was unsure how best to use it to her clients' advantage. There was the possibility of arguing a spoliation-of-evidence theory, or that the corps had demonstrated contempt of Congress. Either way, the scientists' loss of access to the site amounted to a fairly substantial advantage in the litigation.

29

LIE DETECTOR

"Twenty-five million dollars!"

Benton County coroner Floyd Johnson could not fathom any bones being worth that kind of money. But the FBI had told him that the estimated value of the femurs on the black market was in that range. Besides being the oldest complete skeleton ever discovered in North America, Kennewick Man had received tremendous publicity around the world. His bones were potentially a trophy to ancient fossil collectors. The FBI wanted to talk to Johnson about his handling of the bones prior to turning them over to the Army Corps of Engineers.

Johnson asked Benton County prosecutor Andy Miller for advice. Miller called John Schultz, a sixty-year-old criminal defense lawyer who had practiced in the Tri-Cities area since 1964. Schultz knew Johnson well. He had defended some of the accused criminals whom Johnson had arrested as a police officer, prior to becoming the coroner. He agreed to meet with Johnson.

"Floyd, what are we going to be doing here?" Schultz asked. "What's this all about?"

Johnson explained what he knew about the case and indicated that he wasn't afraid to talk to the FBI. "I've got nothing to hide," he said.

After talking with Johnson, Schultz accompanied him to a meeting with the FBI. The interview went smoothly. Johnson openly answered all of their questions.

"I think I'm done now," one of the FBI agents said, looking directly at Johnson. "But I have one more question. Would you be willing to take a polygraph?"

"Yes, I would," Johnson snapped.

The agent raised his eyebrows, then glanced at Schultz before looking back at Johnson. "Do you want to talk to your lawyer about this first?"

Johnson looked at Schultz.

"I'll talk to him about it," Schultz said.

Moments later, Johnson confirmed that he would submit to a lie detector test.

Days later
Federal Building
Richland, Washington

Seated on a soft cushioned chair in front of a rectangular table, Floyd Johnson faced two FBI agents.

"Lift up your arms, please," one of them said.

Johnson raised his arms, and the agent wrapped a belt with a tubular rubber monitoring device on it around Johnson's chest, fastening it together with hooks. He strapped another, similar device around Johnson's waist.

"Go ahead and have a seat," the agent said.

Johnson sat down. The agent attached blood-pressure-monitoring devices to his fingers.

"Please sit straight up and place your hands on your hips," the agent said.

Johnson repositioned himself on the chair.

"State your name, age, and where you were born," the agent said.

Johnson did.

"Now, select a number from one to ten. I'll say each number. When I come to the number you've selected, say 'No.'"

As instructed, Johnson lied when the agent called out the number he had selected, giving the agents an opportunity to record his breathing and blood pressure at that moment.

"Now we just want you to answer yes or no to the following questions."

"OK."

"Have you ever lied in your testimony in court?"

"No."

"Have you ever falsified a report to make yourself look better?"

"No."

The questions continued for nearly three hours.

"Did you take the bones?" the agent finally asked.

"No."

Johnson stood up and began unhooking the monitoring devices. "That's enough," he said.

"We'd like you to answer a few more questions."

"Nope," said Johnson, exhausted. "That's enough."

30

SKIN THICKNESS

Late April 1998
Portland, Oregon

"These things grow over a period of time."

The doctor's explanation hardly comforted Paula Barran and her husband, Richard. They had gone to see the doctor to discuss the biopsy results from a tumor detected on Richard's colon.

"What's going on?" Barran asked.

"Well, it's not frankly malignant," the physician responded.

"What does that mean?"

The physician explained that the tumor didn't appear to be malignant but nonetheless needed to be removed at once. It was six inches long and wrapped around Richard's colon. Surgery was scheduled for May 28th.

After leaving the doctor's office, Barran checked her office calendar. The judge had scheduled the hearing on the disappearing femurs for May 28. She had an obligation to make oral arguments in federal court at the same time her husband had to undergo cancer surgery. She had an important decision to make.

May 27, 1998
Carson City, Nevada

Owsley followed artist Sharon Long into her studio. Paint, brushes, plaster, rubber compound, and sculpting tools cluttered the top of a wooden counter with a wall mirror above it. On the table was a plaster skull covered in clay molded to the shape of the Spirit Cave mummy's cranium and face, mounted atop an eight-inch-high wooden stand.

"OK, here's what we have," Long said, pointing to thirty-three tiny pegs—skin tissue thickness markers made of eraser—protruding from the clay. To figure out the skin thickness of the Spirit Cave mummy, Long relied on a study performed by scientists who measured skin thickness on hundreds of cadavers housed in metropolitan morgues around the United States. Long's eraser pegs marked thirty-three differ-ent thickness points on the human face. "Here's point one," she said, pulling a peg from the clay face.

"How many times did you check that tissue thickness?" asked Owsley, familiar with the study that Long had relied on.

Long said she had triple-checked all the thickness measurements. "Tissue thickness markers don't lie," she said. "They are based on averages."

"Well, the averages are for people today," Owsley pointed out. "And taking into consideration the fact that cadavers are lying flat and flesh falls back when the body is in the prone position, let's take a couple millimeters off here," Owsley said, pointing to the cheek area.

His thoroughness caught her off guard.

Long reached for a sculpting tool. As she thinned out the face, Owsley talked her through a visual image of the face, supporting the image with some photographs of modern-day Ainu, the indigenous people of Japan. To prepare for the facial reconstruction of the Spirit Cave man, Owsley had examined Ainu skeletons at the American Museum of Natural History in New York City and at UCLA. He had studied a half dozen Ainu altogether, confirming Jantz's initial statisti-cal study revealing that the Spirit Cave mummy most resembled the Ainu. Looking at other Ainu had given Owsley a better understanding of their skull size and shape.

Using the wire loop end of the tool to peel away clay, Long narrowed the face as if peeling an apple, listening intently to Owsley. Native Americans have a particular appearance, he pointed out. And people in northern Asia today—Siberians and so forth—look just like that: broad faced with broad crania. "You can see the connection between these people and modern Native Americans," Owsley said.

But the Spirit Cave mummy looked totally different. Owsley insisted that he and Jantz could draw a connection—a line between Spirit Cave and modern Ainu, some nine thousand years—that shows a closer continuity between the Old World and the New World over a broad time span than you can in the same continent through time.

Long didn't doubt Owsley's description of the Spirit Cave mummy. She glanced at Owsley through the mirror, his bushy hair uncombed and his shirt collar wrinkled and open. As if in a large college auditorium, he lectured to an audience of one. She hated to interrupt.

"When you can take a break," she said, "look at this and see if that's enough."

"Hmm, let's see." Owsley examined the entire face. "Well, take a little more. Let's narrow his face down a little bit. Thin him out a *little* more."

"Where do you want me to take it down?"

"Try taking a few millimeters off the fullness of the cheek. Remember, these people were hunter-gatherers."

"They weren't eating baked goods," Long joked, shaving more off the cheeks. "OK. Check this."

"There. That's it. You got it. That's it. Now let's start aging him."

Long reached for a dental pick with a fine-pointed tip.

Slowly she made incisions in the skin, forming wrinkle lines. When she reached the lips, Owsley stopped her.

He wanted her to make the lips more lined and not so smooth looking, like somebody that sits behind a desk all the time. "Make his lips look like they are real dried out from the weather," he said. "These people were in the weather all the time. Their lips and skin would have looked a lot more lined."

Long accentuated the lines in the lips. While she worked, Owsley discussed the Spirit Cave mummy and his lifestyle. This guy was probably forty-five when he died. He had many injuries. He had arthritis. He had deterioration in his spine.

Long asked about the mummy's habitat.

"There were marshes around here," Owsley said. "And they were fishermen. They had nets."

"What do you think, Doug?"

"Deeper," he said, looking at the wrinkles around the mouth region. "There would be deeper wrinkles around the mouth. He was forty to forty-five years old. But you have to make him look like somebody in our time period that is fifty to fifty-five. These people were in weather all the time, like ranchers in Wyoming, out in the wind rounding up cattle."

For the next three days and evenings, Owsley and Long sat in her studio, meticulously going over every millimeter of the facial reconstruction. When she finished the shaping and aging, Long turned on a tiny air compressor hooked up to an airbrush. Holding the brush like a pen, she sprayed flesh-colored acrylic paint evenly over the face.

While it dried, Long asked Owsley about eyelash and hair color. She had ordered a variety of false eyelashes and wigs. "What should we do about the hair?" Long asked. "How long should it be?"

There's an image of what these ancient people looked like, Owsley explained, a Siberian type of person running around with a spear taking on this big flailing mammoth or mastodon. "That's the image you see in the textbooks or in murals," he said.

Long agreed.

"But that look *isn't the look*," said Owsley, explaining that the Ainu, despite being an Asian population, are very different from the so-called Mongoloid peoples like the Japanese and Chinese and Koreans. They have wavy hair, very hairy full beards. They have more of a Caucasian type of appearance. "They're Asian in their origins but it is unclear whether they are an early Caucasoid group that got in there and got isolated," Owsley said.

Long wondered whether she should make his hair dark or light.

Owsley suggested a less descriptive color, something in between, in order to fend off critics who might argue that they had tried to make the mummy appear as if it belonged to a certain race.

Long selected a brown, wavy-haired wig and placed it on the Spirit Cave man's head.

Owsley nodded.

These early people were going into Asia and into Europe, he explained. Drawing an ancient European-Asian connection was natural, they're quite similar because they're coming out of the same ancestral population. They were running into archaic Homo sapiens in Asia and Europe that were there before the Neanderthals. Owsley felt that the Ainu, as newcomers to the area, were more sophisticated culturally and replaced the existing primitive population over time. Later, however, early Japanese rice farmers drove the Ainu to the fringes of Japan. The farmers had large families, while the Ainu had small families. By 1800 or thereabouts, the original distribution of the Ainu was greatly restricted to the northern Hokkaido Island area.

"What about eye color?" Long asked.

"He should have dark eyes," Owsley said.

Once Long set the hair, eyes, and eyelashes, the Spirit Cave man appeared complete.

"That's the look," Owsley said, admiring Long's talent. "You have captured him."

31

OBJECTION

May 28, 1998
Sisters of Providence Hospital
Portland, Oregon

It was 5:30 A.M. when Paula Barran stopped her blue Volvo 850 outside the hospital. Richard Hunt slowly got out. "I'll call and check on you," she said, at a loss for words. Hunt said nothing.

Pulling away, she questioned her ability to compartmentalize and focus on the case while Hunt was in the hospital. She arrived at her law office by six. With three hours until she had to be in court, she closed her office door and immersed herself in her oral argument. If she allowed herself to think about what was going on at the hospital, she would not be able to function in court.

At 8:00 her phone rang. It was Hunt. "I'm checked into my room," he said. "Now it's just hurry up and wait."

"What room are you in?"

He gave her the number and said he would talk to her after the surgery.

"You're going to be OK," Barran said.

After she hung up, she took a deep breath, gathered her files together, and walked to the courthouse.

U.S. District Court
Portland, Oregon

"The primary issue this morning is the curation of the remains in question," Judge Jelderks said, welcoming Justice Department lawyers and Paula Barran and Alan Schneider to court. He invited Barran to speak first. She had some surprising news for the court. Schneider's investigation into the missing femurs revealed that a high-ranking official at the Battelle facility—Darby Stapp—was married to Julie Longenecker, the Umatilla tribe's archaeologist, who had conducted the inventory of Kennewick Man five days after the skeleton arrived at the facility. "Mr. Stapp is one of the custodians . . . keyed to the vault room," Barran said. "It is also a fact that Mr. Stapp and his spouse have published articles which are highly critical of the scientists' positions in this lawsuit. We have been worried about ongoing security breaches over the curation of this skeleton and now we learn of a relationship between highly placed people which doesn't seem to be supervised by the curator."

Jelderks occasionally jotted down notes as Barran revealed that Schneider had also confirmed through discovery documents that members of the Indian tribes had gotten access to Kennewick Man to perform a religious ceremony. The tribes had also gained access to the storage vault on another occasion and grabbed other bones, including rib and vertebra fragments believed to be from Kennewick Man. The tribal members had buried those bones at an undisclosed location. "We don't know why the tribal claimants were not kept out of the vault," Barran said. "We don't know where these bones were reburied."

"I think you can draw some inferences from this," Barran continued. "The first is that from the day this skeleton has come into the custody of the Army Corps of Engineers it has been bungled." She reminded Jelderks that Jim Chatters had performed a complete inventory of Kennewick Man before surrendering him to the corps, and that inventory included the presence of femurs.

"Femur bones are very important bones to an anthropologist trying to decide the origin of this particular skeleton," she said. "They were present when Dr. Chatters had the bones. They were present when he videotaped the packing. They were present when he turned them over to the Benton County coroner. They were in the Benton County coroner's lockup and transferred to the government. And it would not be until March 1998 that the government came back to you and reported that these bones were missing. The government had the bones for 14 full months without doing a formal inventory. There has been no explanation of where those bones are, except for a lot of finger pointing and some suggestion that perhaps Dr. Chatters is a bone thief."

Barran shuffled some papers, pulling from her file an affidavit. "Lieutenant Colonel Curtis, who is the official custodian of these bones right now, says that during the religious ceremonies nothing happened that harmed the bones. But nowhere in any of his many declarations have I ever heard him say that he actually attended those ceremonies. If he wasn't there, I would suggest that he's not exactly a competent witness to tell you that the bones weren't stolen during one of the religious ceremonies."

Speaking rapidly, Barran began to close. "Quite frankly, I believe and I submit that the government has lost all credibility in their ability to handle this skeleton and the time is now for the court to intervene and require the skeleton to be preserved by a neutral curator who answers only to the court."

Schneider's investigation did not lead to who had the bones or whether they were even stolen. But he had uncovered a glaring conflict of interest for a ranking member of the Battelle facility who had direct ties to the Umatilla tribe. It was clear that the tribes had been getting unlimited access to Kennewick Man, despite a court order to keep him under lock and key. Besides the missing femurs, other bones had been removed from the facility. To Barran it was clear: the facility was not a safe place for Kennewick Man.

"I'm very concerned about personal relationships between Battelle employees and the claimants. I'm concerned about inadequate security. I'm concerned about what I call the very serious coincidences that the tribal claimants somehow knew how to get into the vault room and then somehow knew what box to take. I'm concerned that there's been no effort to recover what's been taken."

Barran stopped. "Thank you, Your Honor."

Jelderks looked toward the table of government lawyers. Unable to close out Barran and Schneider sooner, the Justice Department had reassigned a new lead attorney to the case, Robin Michael. She began by rehearsing the chronology leading up to the discovery of the missing femurs, pointing out that on September 5, 1996, the skeleton arrived at Battelle in a wooden evidence box that was sealed with tape. "They remained in that manner until September 10th when the Corps archaeologist opened the box for the first time and conducted an inventory in the presence of two other staff archaeologists," Michael said. "That inventory indicated that there were missing femur fragments. At the time the inventory was taken the remains were in plastic Ziploc bags in the manner that they had been provided, transferred by Dr. Chatters. They were not taken out of their bags when the inventory was done. They have not been handled by human hands since the remains have been transferred to the Corps.

"We have no accounting in regards to what handling of the remains occurred prior to Dr. Chatters turning over the remains."

As Michaels wrapped up, Barran sipped her glass of water.

"Ms. Barran, you may call your first witness."

Barran called Chatters.

Chatters took a seat in front of the microphone.

"Are you the person who initially collected the Kennewick Man remains?" Barran asked.

"The complete skeleton, yes."

"Did you have possession of femur bones when you had the Kennewick Man remains in your possession?"

"Yes, I did."

"What did you do with them?"

"Placed them in plastic bags, individual fragments, I believe, in each bag and placed them in the bottom of the wooden box."

As Barran continued, one of the U.S. attorneys interrupted her. "Your Honor," he said, "we need to either take a sidebar or discuss an issue here in Court about this particular testimony." The U.S. attorney informed Judge Jelderks that Chatters had a criminal attorney and had exercised his right not to talk to the FBI about the missing femurs. "I'm assuming he's waiving his rights then at this point," the U.S. attorney said.

"I don't think that's for me to decide," Jelderks said.

"Okay," the U.S. attorney said.

Barran resumed her questioning of Chatters, then sat down.

Jelderks looked at Robin Michael. Her cross-examination of Chatters focused on the discrepancy between his testimony that he put the femurs in the box and the inventory done at Battelle, which indicated that the femur fragments were absent. "You're aware of the fact that the inventory that was conducted at that time showed three to four missing femur fragments; is that correct?"

"I'm not aware of what the inventory said."

"You had no control over the remains between August 30th and September 5th?"

"That's correct."

"So you cannot testify to a hundred percent certainty as to what happened to the remains or to the femur fragments during the period from August 30th and September 5th?"

"That's correct."

"I have nothing further, Your Honor."

"Dr. Chatters, one follow-up question," Barran said. "What were you told by the Benton County Coroner when you were advised that he needed to give up possession of the remains?"

"Objection," Michael said. "Hearsay."

Jelderks looked at Michael. "Objection's overruled."

Chatters looked at Jelderks, making sure it was all right to speak. "That he had been ordered by the county attorney and that the attorney had made agreement with the Corps of Engineers that no further work of any kind would be done on the skeleton and that he had to come get the bones immediately and would be there shortly."

After hearing from Chatters, the government called its witness. While he testified, Barran suddenly looked up at her watch. It was 11 A.M. A wave of panic suddenly swept over her. Wondering what was happening to Hunt, she momentarily lost her train of thought before convincing herself not to think about it until 1:00.

When the government's witness finished, Jelderks adjourned. He promised a swift decision on whether to order the skeleton transferred to another, more neutral facility. As soon as Barran got back to her office, she called the hospital. A nurse informed her that Hunt was out of surgery and back in his room resting comfortably. "He's pretty groggy," the nurse said.

The next day, Judge Jelderks issued an order. "Defendants shall provide Dr. Owsley access to the skeleton to evaluate the condition of curation," Jelderks wrote, ordering the government to cooperate with Owsley in arranging a time and procedure for Owsley to see Kennewick Man.

Schneider and Barran wanted Chatters to accompany Owsley. He was the only one who had examined the skeleton previously and was therefore in a better position to compare its current condition with its condition at the time Chatters surrendered it to the corps. The government adamantly opposed Chatters's participation. The judge resolved the impasse with another order on June 12.

"Both Dr. Owsley and Dr. Chatters may participate in the evaluation of the condition of curation," Jelderks ordered. He also ordered that a complete inventory of Kennewick Man's bones be conducted prior to the skeleton's being transferred from Battelle to the Burke Museum in Seattle. The judge said that if both sides agreed, the inventory could take place simultaneously with Owsley's inspection of Kennewick Man. He said that if the government refused to have the inventory done while Owlsey was there, the plaintiffs had the right to send two representatives to be present when the inventory was conducted.

32

ONE LOOK

U.S. Attorney Robin Michael and Alan Schneider had negotiated the final terms governing the examination and inventory of Kennewick Man by Owsley. By mid-September, they had drafted a joint memorandum of understanding, spelling out the terms. Owsley would be permitted one personal assistant and the Army Corps would supervise the inventory. The details were vague. Schneider wanted more specific guarantees that the corps would not impose any last-minute restrictions on Owsley that would crimp his standard method of operation for inventorying the bones. "This is as good as it gets," Michael said, insisting that the restrictions that Owsley and his colleagues had encountered in the past were initiated by her predecessors. She assured Schneider that she had cleaned house and brought in a new team of lawyers with a different approach to the case. "Trust us."

Schneider didn't, but he reluctantly signed the joint memorandum of understanding with the Justice Department. Days later, Owsley spoke by telephone with a curator from the corps who was assigned to monitor Owsley. Together they ironed out specifics of the protocol for the inventory, including the names of the individuals who would be present and their specific roles. Owsley followed up the phone call with a letter to confirm their verbal agreement. "This letter follows our phone conversation of a few weeks ago," he began, acknowledging that numerous government-appointed conservators would be on hand to assist, as well as tribal officials to observe. "With this many people in a small space for an extended time, the observers probably should wear masks to reduce the possibility of airborne DNA contamination. It would be good to check with a molecular biologist and with the conservators to assess the risk of further contamination."

Owsley asked if the corps would supply masks. "I'll bring a lab coat, to wear, one sliding caliper, the accepted forms, and a laptop," Owsley wrote. "The software for the laptop is WordPerfect. Thanks for allowing me to participate in the inventory. Best wishes."

Days after Owsley sent his letter, the government notified Schneider that Owsley would not be permitted to bring a laptop inside the facility. And they weren't going to let Owsley's designated personal assistant, Cleone Hawkinson, accompany him unless Jim Chatters got left behind. They would not let more than two representatives from the plaintiffs enter the facility. I'm getting really pissed off, thought Schneider. From the moment we said we wanted Chatters to participate in this process, the corps has been waiting for some way to hit us in the chops.

Judge Jelderks resolved the dispute on October 20. "Pursuant to this court's order of June 12, 1998, plaintiffs may have two representatives present during the exit inventory," he ordered. "Dr. Chatters is not disqualified from being one of those representatives. If Ms. Hawkinson is present, she counts as one representative. Dr. Owsley may use an audio recording device to take notes in lieu of Ms. Hawkinson."

Forced to choose between Chatters and Hawkinson, Owsley chose Chatters, opting to use a tape recorder rather than a computer to record the inventory results.

Wednesday, October 27, 1998
Alan Schneider's law office
Portland, Oregon

To Schneider, the corps was practicing petty politics by denying Owsley access to his laptop in the inventory process. It was just another attempt to make things difficult.

Seated beside Owsley, Hawkinson had a package of blank ninety-minute tapes that she had prelabeled. She had also rented a high-performance tape recorder with a clip-on microphone for Owsley to wear during the inventory. Owsley had never done an inventory this detailed without an assistant to record. Hawkinson hoped to make the new experience as easy as possible.

"At this point, the government is playing absolute hardball with us," Schneider said. "So there are a few things I want to go over with you about tomorrow."

He told Owsley to document everything, but he insisted that Owsley keep all of his observations to himself, particularly any unusual features or trauma in the bones. He coached Owsley to be discreet and limit what he said on tape. The less Owsley said, the less opportunity the government would have to look for inconsistencies in his statements.

Owsley's problem was that he liked to identify unusual characteristics in bones and hypothesize on potential explanations until he figured out the cause. That's what scientists do.

He told Owsley to be sure to keep the audiotapes, reminding him that they were his personal property and the government had no right to ask for the originals.

"Just remember, you're going into an environment where the people are not your friends," Schneider said to Owsley and Hawkinson. "There's nobody up there that's going to be our friend except the two of you and Chatters."

On the 225-mile drive to Richland, Washington, Owsley sat in the passenger seat of Hawkinson's car, reviewing the bones in an articulated foot. He didn't normally work with feet, and he didn't want to make any mistakes on Kennewick Man. He knew he could end up being the only scientist to handle Kennewick Man, making his bone inventory the only source for future scientists to rely on. Yet he would be racing the clock to identify roughly three hundred bone fragments. And he would have no laptop and no personal assistant, only the watchful eye of an adversarial team.

October 28

6:45 A.M.

Battelle Laboratory

As sunlight peeked over the horizon, Hawkinson parked her car outside the Battelle facility. Bottles of orange juice and cranberry-apple juice tucked under his arm, Owsley stepped out. With his free hand, he grabbed his backpack and the case with the tape-recording device. "Call me when you're through if you need a ride," she said. "Otherwise I'll see you when you get back to the motel."

Wearing a windbreaker over his dark blue tie and collared shirt, he approached a set of glass doors bearing a sign: "Identification badges are required and must be worn in this building." He entered and signed in.

Inside, Owsley encountered three conservators hired by the corps, tribal members, and Justice Department lawyers who looked on as Owsley opened his briefcase and pulled out six perfectly sharpened pencils and clipped them to his shirt pocket. He plugged in his tape-recording device and inserted a blank ninety-minute tape labeled "#1." Then he clipped the microphone to his shirt.

"I'm ready," he said.

At 7:00 A.M. a technician brought out rectangular plastic storage boxes with lids. Two of the corps' anthropologists, Rhonda Lueck and Teresa "Terry" Militello, both of whom studied under Owsley previously, stepped forward. Terry and Rhonda began removing Ziploc bags from the container. Each bag contained groups of bones, and had an identification number taped to the outside. The individual bones did not have identification numbers. Over the next two hours, Terry and Rhonda removed the bones from each bag and placed them on metal

trays that had sheets of paper with numbers corresponding to the numbers on the bags. Owsley put on a pair of latex gloves and began indi-
vidually identifying approximately 380 bones and fragments taken from
the bags.

He handed Rhonda a form—the words "Coding Instructions for
Skeletal Inventory" appeared across the top—and explained how to code
each bone on the form. "I'm going to start a series of small fragmented
pieces of bone that is in the 'unidentified' category," he said. "OK?"

Rhonda nodded her head.

One by one, he picked up unmarked fragments, piecing them
together and identifying them. Concentrating, Owsley worked through
the bones as if he were alone in the room. Plodding through tray after
tray, he never paused except to change tapes every ninety minutes. It
was 2:30 by the time he identified the last unidentified fragment.

Stopping only momentarily to call Chatters and notify him that he
was moving on to the larger, identified bones, Owsley started first with
the skull. He immediately observed its weight. "The cranial vault is
heavy," he began, holding it up to the light to get a better look inside it.
"It has soil inside the vault, especially in the frontal area, especially in
the upper nasal area. In the cranial surface there is caked dirt. The
largest quantity of soil available for this individual is found inside the
cranium."

Cradling the cranium in his hands, he examined the suture lines
running through the skull. They were closed. Owsley figured Kennewick Man had to have been middle-aged when he died.

He rotated the skull in his hands. A huge fracture on the side of the
vault intrigued him. He wondered what caused it and itched to study it
more closely. But he only had time to scan. He observed a much more
subtle fracture line along the right side of the skull. He tried to determine whether it had occurred before or after death. "It has postmortem
breakage," he concluded, "relatively recent breakage a little lighter in
the fracture margin. It has soil embedded."

Carefully studying the base of the skull, he spotted a depression in the
bone, signaling to him that the bone had previously been injured, then
healed. He measured the healed area with his calipers. Uninterested in

Kennewick Man's injuries, Owsley's observers instead studied him, frowning stoically as he took extra time to interpret the skull's pathology. Oblivious of their stares, Owsley instinctively noted Kennewick Man's sloped forehead and slightly developed brow ridges, classic male features. But the extremely narrow frontal bone struck Owsley as unusual.

Owsley looked at the rim of the right eye orbit. Then he picked up a small piece that appeared to fit there. "With the right malar, these two attach together to form a complete right eye orbit," he said as he started reconstructing the face by fitting miscellaneous fragments from the table to the cranium. "It will articulate and match perfectly with the right malar and also joins up with the nose."

Kennewick Man's face started to take shape. "There is a depression, an oval-shaped depression, in the frontal area of the right malar, and we'll need to think about what significance that is."

"How should I score that?" Rhonda asked.

"Say inferior orbital margin," Doug said, never looking away from the cranium.

Suddenly, Jim Chatters entered the lab. After being introduced to the curators, Chatters sat opposite Owsley. Relying on the black-and-white photographs he had previously taken of the skeleton, along with some hand sketches he had drawn of the skull, Chatters analyzed the existing fractures to see if they had expanded or whether new fractures had surfaced. He noticed two hairline cracks, one above each eye socket. The cracks had not been there when the skeleton left his house two years earlier.

Then he examined the cracks in the skull that had been there before, observing that they had enlarged and expanded. "Doug, can I borrow your calipers?"

Without looking up, Owsley passed him the calipers.

"You can't do that," Rhonda interjected, looking directly at Chatters. Chatters froze.

"You can't take any data!" Rhonda said.

"Look, I'm here to assess the condition of the skeleton," Chatters said. "I can't do that unless I measure the cracks."

Owsley stopped working. "He's not going to make any measurements of research quality," he said calmly. "He's just trying to measure to see what the condition of the skeleton is."

As Chatters confirmed that the width of the cracks on the left side of the cranium had doubled since being in the federal government's custody, Owsley started looking at Kennewick Man's teeth. A lot of the facial surface enamel had worn away, as had most of the crowns. Then he observed a tooth that appeared to have been abscessing at the time of Kennewick Man's death. The roots were clearly visible.

At 6:30 the group inside the laboratory insisted on breaking for dinner. Owsley, who tended to forget about eating and drinking while working, agreed. After the break, he reattached his microphone. Energized and appearing youthful, Owsley reached for the proximal half of the left femur, the only fragment remaining since the disappearance of the femur parts. He fully extended his caliper. "The caliper isn't long enough to measure the length exactly, but it's approximately two hundred twenty-two millimeters in length," he said. "The left femur has a transverse fracture, postmortem with a somewhat lighter fracture margin, indicating relatively recent postmortem damage."

Working through the fragments took hours. Fatigued and struggling to concentrate, the curators yearned to call it a day. Owsley had no intention of stopping. The corps was limiting him to one visit. He wasn't leaving until he had gone over every last bone fragment in the collection.

Doug turned to Chatters. "What else have I not seen? You have hands taken care of?"

Chatters nodded his head affirmatively.

"You have hands and feet taken care of?"

"Yes."

"You have it listed in a way that there's a record as to what there is?"

Chatters nodded.

The bones were all accounted for. Only the preliminary taphonomy examination remained, an evaluation for clues about the position of the body after death. Determining whether the body was flexed, flat, or faceup or -down was critical for establishing whether Kennewick Man had received an intentional cultural burial or had fallen victim to an accidental burial, such as from a mud slide, landslide, or homicide.

It was now past midnight. Owsley had been at the facility for nearly eighteen straight hours. Everyone else was struggling to keep their eyes open. But Owsley had a penchant for working through the night and going without sleep when engrossed in his work. Back in college he had developed unusual sleep patterns that enabled him to study for twenty-four hours without sleep. Or he would get by on five hours' sleep per day when he would engage in marathon or "binge" studying, a process where he would study for two or three straight days on little sleep. Since coming to the Smithsonian, he had taken this practice to another level. It was routine for him not to leave his laboratory for two or three days. Instead of going home to sleep at night, he would catnap on the floor in his office. Sleep was the enemy whenever he was engrossed in studying a skeleton.

To the dismay of the corps officials, Owsley took no account of the lateness of the hour. He examined the bone color, looking for evidence of stress. Most of the tones were tan to light brown. Some of the bones had a greenish tone, probably due to algae discoloration from being in the water.

Owsley paused. "I saw no well-defined rodent tooth marks. I saw no significant carnivore tooth marks. I did see evidence of plant rootlets in a few of the vertebrae. But there's no obvious damage due to root etching.

"And I saw no warpage due to ground pressure. There's no evidence of burning."

He had just two categories left: cultural modification and adherent materials.

"I saw no cut marks; no intentional fracturing; no evidence of postmortem drilling, cutting, or other modifications. As far as adherent materials go, there is no desiccated tissue; no textile impressions; no hair or fur; no unknown material associated with it."

Doug looked at his watch. "It is 2:30 A.M. on October 29," he said. "This concludes this analysis at 2:30 A.M. The conservators are still working to get things packed up."

Owsley stopped the recording device, removed the microphone

from his shirt, and helped repack the Kennewick Man bones into containers. At 4:12 A.M., twenty-one and a half hours after starting, he signed the log and exited the facility. Three conservators from the facility drove him back to his hotel.

After three hours of sleep and talking with a local reporter about Kennewick Man, Owsley rode with Hawkinson to the discovery site. She parked facing the Columbia River in a lot that held only one other car and a dented garbage can chained to a post. Following directions they had received from Chatters, they searched for the area along the bank where Kennewick Man surfaced.

"Jim said there should be a track through there," Hawkinson said, pointing at a mass of tangled limbs, tall brush, and blackberry brambles.

"I don't think a rabbit could get through that," Owsley said, suggesting they go back up to the road and look for another way in.

Owsley led Hawkinson along the road, searching for a less overgrown area to pass through to get to the shore. Pushing branches from his path and climbing over thick brush, Owsley trudged along. Soon they came upon a bunch of spindly Russian olive trees, planted by the corps the previous May. In five months they had grown to over six feet in height.

Owsley looked down. Fiber matting emerged through places in the soil, covering the rocks that the corps had poured over the area. The rocks and matting stretched toward the shore, where the Columbia River lapped against the rocks. "The corps made sure nobody would get at this place again," Owsley said, determined to search the area. "I want to take a closer look in that grove of trees along the curve to that point." It's secluded and shielded, perfect forensic conditions, he thought.

He stooped over, his eyes focused on the ground. It would have been great, he thought, to find more evidence of Kennewick Man's contemporaries. There had to have been more evidence to find. For Owsley, there was always more evidence; it was just a matter of digging. But this time there would be no digging for more clues. He wished they could somehow have stopped the corps.

An hour and a half later, Owsley emerged, his hands gripping a collection of empty beer cans and other litter. "They even went to the trouble to put in a sprinkler system," Owsley said to Hawkinson, lamenting the cover-up of the site by the corps as he deposited the litter in a nearby trash can. "They were very thorough."

Before Owsley and Hawkinson returned to Portland, the Kennewick Man had arrived at the Burke Museum in Seattle. Under court order, he was placed in a curation room, out of public view and off limits to anyone not authorized by Judge Jelderks. Over the next year, a team of government-selected and court-approved scientists studied the skeleton in secrecy, attempting to determine if it was affiliated to the Umatilla and other tribes that had claimed him. The judge agreed to delay proceeding until their report was complete.

When the results came in, they confirmed what Owsley had been predicting all along. Kennewick Man did not resemble modern American Indians. He looked much more like the Ainu.

33

GOING DEEP

February 1999
Museum of Natural History
Washington, D.C.

Good science is often too broad to fit through narrow minds. The more Dennis Stanford studied the traditional Clovis model—the New World was first populated by people who came from northern Asia over the land bridge and migrated down to present-day America, ultimately giving rise to Native Americans—the more he concluded the traditional thinking was far too simplified. Since archaeologists first uncovered the original Clovis site in New Mexico, roughly forty others had surfaced in North America. Stanford had visited every one. What he found convinced him that Clovis people did not come from northern Asia and didn't enter the New World over the land bridge. The age and location of the sites seemed to confirm his conviction.

The oldest Clovis site, outside Nashville, Tennessee, dates back 14,000 years. The next oldest site is located near Dallas, Texas, and is 13,500 years old. These two, like the others in the eastern half of the United States, predate the original Clovis site in New Mexico and the subsequent ones found in the western states and Mexico. This suggests that the Clovis people started in the East and pushed westward, not the other way around. Also, no Clovis sites have surfaced in Canada, casting more doubt on the likelihood that Clovis people migrated down from Alaska. There are no Clovis sites between the Bering Strait and the modern-day United States.

Stanford also looked at the style of stone tools and projectile points turning up consistently at Clovis sites. They were inconsistent with those being produced in northern Asia at the time the land bridge was open. Rather, the artifacts matched those being produced in northwestern Spain as far back as eighteen thousand years ago. Stanford believed the Clovis people were not the first ones to populate the New World, but instead developed from a preexisting culture, one that had ties to the Solutrean, a prehistoric culture from Europe. Unwilling to deny that there might have been some gene flow from Clovis people to modern Native Americans, Stanford remained unsure how or if Clovis related to American Indians. He was convinced, however, that Clovis people were not alone in the New World.

Owsley agreed. Kennewick Man, the Spirit Cave mummy, and other recently examined skeletons in the West were producing a profile of other human populations, possibly from the Asian rim. Yet these people were not quite old enough to have crossed the land bridge at the Bering Strait, as the closing of the bridge predated the age of Kennewick Man. Did Clovis people look more like modern-day Native Americans or like the Pacific rim populations that seemed to be reflected in Kennewick Man and the Spirit Cave mummy?

The answer was elusive. The missing element in all the Clovis sites was human remains. No skeletons of the Clovis people had ever turned up. But shortly after Kennewick Man, Stanford got an idea. Back in 1973 he had examined the largest cache of Clovis artifacts ever found. They came from a site in Wilsall, Montana, a small town situated in an intermountain basin between the Crazy Mountains and the Bridger Mountains. The site was called the Anzick site, after the Anzicks, a family that owned the property on which the artifacts were found.

He decided to tell Owsley about the site and fill him in on the background.

In 1968 a construction worker operating a front-end loader was removing soil from a sandstone outcrop on the Anzick property when he spotted a stone tool. The worker discontinued digging and entered the excavated area, approximately eight feet belowground. He and some friends found a three-foot-by-three-foot pit containing stone tools and projectile points stacked like a deck of cards. Human bones were buried beneath the artifacts. In all, the workers recovered roughly 125 stone tools, fluted projectiles, a half dozen bone tools, nonhuman bones, and a few human bones, including skull fragments.

Shortly after the discovery, a University of Montana archaeologist examined the site and the cache it contained. In a report published in 1969, the archaeologist confirmed that the artifacts were consistent with the Clovis culture, but lamented the inability to link the human remains to the artifacts. "Had it been possible to make a definite association of the human bones with Clovis materials," the report read, "it would have given archaeologists their first glimpse of the actual bones of one of these ancient hunters."

At that time, molecular-level radiocarbon-dating techniques were not available. Nor could archaeologists date the remains by the fact that they were recovered with the Clovis artifacts. The amateur methods of recovery used by the construction worker and his associates made it impossible to tell whether the bones and the artifacts were deposited at the same depth beneath the earth's surface.

After the discovery, the cache, including the human remains, were kept by a collector. Larry Lahen, a graduate student at the University of Montana at the time of the discovery who reinvestigated the discovery site in 1971, kept tabs on the cache's whereabouts over the decades that followed. He eventually gathered the materials together on loan and had them displayed at the Montana Historical Society Museum. Then in 1997, the human bones were finally radiocarbon-dated with the most up-to-date technology. The skull fragments turned out to be slightly over 12,600 years old.

If the bones were indeed the remains of a Clovis person, they would help date the time that Clovis people were in the Montana region as roughly 1,400 years after they were in the Tennessee area. This would support Stanford's hypothesis that Clovis people migrated westward.

After talking with Stanford, Owsley wanted to see the bones at once.

He contacted Larry Lahen, who informed him that the skeletal fragments seemed to represent two individuals. But a lot of confusion surrounded their identity. In the thirty-plus years since the discovery, no osteologist had examined the remains. Owsley made arrangements to fly to Montana and examine them.

Most of the bone fragments were cranial pieces. Owsley confirmed that they came from two individuals. One was an infant less than two years old. The other was approximately seven years old. The remains of the infant consisted of twenty-eight skull fragments, and portions of the clavicle and three ribs. All of the infant bones were stained with red ocher, a mineral oxide of iron found in sandy clay soil. The tools from the cache were also ocher-stained, suggesting that the infant and the cache were buried simultaneously.

The skull fragments from the seven-year-old did not have red ocher stain. And further radiocarbon-dating confirmed that the fragments from the seven-year-old were two thousand years more recent. Interviews with the original discoverers revealed that the remains of the seven-year-old were not found in the same pit as the Clovis cache, but surfaced nearby in a subsequent visit to the site while the discoverers searched for more artifacts.

With so few bones from either child remaining, Owsley could draw no conclusion as to their cause of death. The skull fragments of the seven-year-old were also too sparse to measure or draw conclusions from. But the infant cranium was more complete. Owsley held the cranial vault in his hand and observed that it was very different from the scores of infant crania of Native Americans he had studied over the years. Rather than a short, round cranium, the Clovis infant had a long, narrow vault.

Owsley was not ready to say which prehistoric human population the Anzick infant belonged to. But Stanford confirmed that the artifacts buried with it were from the tradition of those found in Europe, not northern Asia. The Clovis people might not be related to Native Americans, or to the human populations that Kennewick Man and the Spirit Cave mummy derived from. The questions were tantalizing. But both Stanford and Owsley agreed: the evidence was building that a host of different human populations were getting into the New World earlier than previously suggested and through multiple routes.

34

SPIN CYCLE

August 15, 2000
Reno, Nevada

"Ancient human remains from Spirit Cave are Native American but they cannot be culturally affiliated with the Fallon Paiute–Shoshone Tribe, or with any other contemporary group."

The announcement from the federal government's Bureau of Land Management surprised Paula Barran and Alan Schneider. They had expected the government to declare the mummy Native American, as it had done with Kennewick Man. "Spirit Cave Man will remain in federal ownership," the announcement concluded.

"The probability is high that the government is going to say Kennewick Man is not culturally affiliated to the tribes in our case," Barran told Schneider.

Schneider agreed. Every argument he could come up with—scientific, legal, or otherwise—convinced him that Kennewick Man could not be deemed an ancestor of the contemporary Umatilla Indian tribe. Roughly 440 generations separated them. If any marriage outside the tribe took place during that span, the amount of genes passed to the next generation was cut in half. Schneider figured that if you cut something in half 440 times, the contribution of DNA would be comparable to a blade of grass in an entire football field.

The Spirit Cave decision encouraged Barran. It also got her thinking. With Kennewick Man virtually as removed from present-day populations as the Spirit Cave man, a consistent ruling by the federal government would deem it also unaffiliated to modern tribes. She wondered
how such a decision would impact her case. What would happen to the
lawsuit?

Interior Secretary Bruce Babbitt, meanwhile, faced a tough political
decision. Two agencies within his department disagreed vehemently
over the future of Kennewick Man. The National Park Service and its
chief archaeologist, Francis McManamon, felt more scientific testing
was in order. But Babbitt's undersecretary, Kevin Gover, who headed
up the Bureau of Indian Affairs, argued that further testing was offen-
sive and unnecessary. Gover told the *Washington Post* that he "couldn't
conceive of what possibly could be learned and why it would be significant" to study Kennewick Man.

Prior to taking over at the BIA, Gover, a lawyer and a Pawnee
Indian, had established himself as an influential fund-raiser and orga-
nizer for President Clinton. After casinos were legalized on Indian
reservations in 1988, the tribes with lucrative casinos became powerful
donors. In 1994, the Umatilla opened a casino resort and eighteen-hole
golf course. With the revenue the tribe hired lawyers and lobbyists,
enabling them to establish a presence in Washington.

The political stakes for Babbitt were high. Native American tribes
had become big financial backers of President Clinton and the
Democrats, and the upcoming presidential election was looming. Fur-
thermore, the results from the government-sponsored study of Kennewick Man had hemmed Babbitt in. In order to deem Kennewick Man
a Native American and repatriate him to the five tribes that had
claimed him, Babbitt had to conclude that by a preponderance of evidence—more likely than not—a cultural link existed between the tribes
and Kennewick Man. Although preponderance of evidence is the lowest standard of proof in law, a degree of evidence that is more in favor
than against is nonetheless still required.

Judge Jelderks imposed a fall 2000 deadline for Babbitt to decide.
Politically, the timing of the deadline could not have been worse.

September 25, 2000
Portland, Oregon

It was 1:00 P.M. when Alan Schneider's secretary informed him that Richard Hill, a reporter from the *Oregonian*, was on the phone. Since the start of the Kennewick Man litigation, Schneider had come to know Hill well. He had been covering the case from the beginning. Schneider had his secretary put the call through.

"Have you heard about the government's decision?" Hill began.

"No. Have they reached a decision?"

"Well, apparently so."

"What is it?"

Hill explained that earlier in the day the Department of the Interior had faxed him a series of documents: a September 25 press release announcing the Interior Department's determination to repatriate Kennewick Man to the Indian tribes; a seven-page letter dated September 21 from Interior Secretary Bruce Babbitt to the secretary of the army, explaining his decision to repatriate; and a forty-page report. Hill told Schneider that Babbitt had concluded that the tribes were culturally affiliated with Kennewick Man on the basis of geography and oral tradition.

This is classic government action, thought Schneider as he listened to Hill while scribbling the words "Geography" and "Oral tradition" on his notepad.

Hill wanted a comment from Schneider.

Schneider wanted to see the documents first.

At 1:11, Hill faxed Schneider a copy of the press release from Babbitt's office and the seven-page letter Babbitt had written to the army secretary. Schneider saw the heading on the Interior Department's press release first: "INTERIOR DEPARTMENT DETERMINES 'KENNEWICK MAN' REMAINS TO GO TO FIVE INDIAN TRIBES." Schneider smirked at the subheading: "Determination by Interior Secretary follows two years of scientific research."

Schneider pulled the page from the machine and started reading.

"The Department completed a careful, detailed series of scientific investigations involving world-class experts to learn as much as possible," the release quoted Babbitt as saying. "After evaluating this complex situation, I believe that it is reasonable to determine that the Kennewick Man remains should be transferred to the Tribes; tribes that have inhabited, hunted and fished this area around the confluence of the Columbia and the Snake Rivers for millennia."

Schneider shook his head, thinking that if the government really wanted to learn as much as possible about Kennewick Man, it would let Owsley and his peers study the skeleton.

"Although ambiguities in the data made this a close call, I was persuaded by the geographic data and oral histories of the five tribes," Babbitt continued in the press release. "If the remains had been 3,000 years old, there would be little debate over whether Kennewick Man was the ancestor of the Upper Plateau Tribes. The line back to 9,000 years, with relatively little evidence in between, made the cultural affiliation determination difficult."

The leap Babbitt had made stunned Schneider. The secretary admitted that if Kennewick Man had been only three thousand years old it would have been easier to link him to the tribes. But Kennewick Man was nine thousand eight hundred years old, and Babbitt admitted that there was virtually no evidence to close the six-thousand-year gap. Yet he decided the skeleton should go to the tribes anyway, citing geography and the tribes' oral traditions as the reasons.

Schneider shared the news with Owsley. From a political standpoint, Babbitt's decision did not surprise Owsley. The tribes had political influence. The scientists did not. Ruling against the tribes would assure a backlash at the Interior Department.

Owsley relished the chance to challenge Babbitt's decision in court. Schneider did too. After notifying Owsley, he called Barran at her office across town.

"Paula, they're gonna give it to the Indians."

"No shit."

"No, I'm serious," he snickered.

"You *gotta* be shittin' me."

"The decision's out. I've got a copy of it right here. I'm going to fax it to you right now."

Amazed, Barran said nothing while Schneider highlighted Babbitt's conclusions.

"The court told the government in 1997 that they can't affiliate the skeleton to the tribes on the basis of geography and oral tradition," he said. "And Babbitt has come back and done it anyway, in effect telling the court, 'We don't care what you said in ninety-seven.'"

Barran was speechless.

"What do you think we ought to do?" Schneider asked.

"Well, let's request a status conference with the judge."

"How should we do that? Should we do that by a phone call? Or do you want to file something?"

"Let's file something so that we can clearly control or at least have influence on the issues that are going to be discussed."

Agreeing to reconvene after Schneider faxed the decision to her, Barran hung up. Her phone rang instantly. A reporter wanted her comment on the government's decision to award Kennewick Man to the tribes. "I haven't seen anything yet, because consistent with what's happened throughout this case, the government has decided to tell everybody in the world before they tell us."

Offering no comment, Barran hung up and retrieved from her fax machine a copy of Babbitt's letter to the secretary of the army. At her desk she skimmed it in search of legal arguments she could make to challenge the decision.

"While some gaps regarding continuity are present," Babbitt wrote, "DOI finds that, in this specific case, the geographic and oral tradition evidence establishes a reasonable link between these remains and the present-day Indian tribe claimants. Cultural affiliation is defined as 'a relationship of shared group identity that may be reasonably traced historically or pre-historically between a present-day Indian tribe . . . and an identifiably earlier group.'"

Barran stopped reading. She knew Babbitt was wrong. But how was she going to prove it?

She looked back at the letter, which went on to explain that the Umatilla and other tribes that had claimed Kennewick Man met the definition of a present-day Indian tribe. Babbitt said that evidence existed linking the cultural characteristics of the present tribes with "the group that lived in the Columbia Plateau region during the life-time of the Kennewick Man."

Barran scratched her head. If there was a group that had lived in the area at the time of the Kennewick Man, she wondered where the bones of the other members of the group were.

She read on. "The oral tradition evidence reveals that the claimant Indian tribes possess similar traditional histories that relate to the Columbia Plateau's past landscape," Babbitt wrote. "After considering and weighing the totality of the circumstances and evidence, DOI has determined that the evidence of cultural continuity is sufficient to show by a preponderance of the evidence that the Kennewick Man remains are culturally affiliated with the present-day Indian tribe claimants."

Barran zeroed in on four words: "preponderance of the evidence." Preponderance means outweighing. In law, it represents the lowest standard of proof, merely requiring evidence that the fact sought to be proved is more probable than not. Babbitt had concluded that more likely than not, Kennewick Man was culturally affiliated with the Umatilla and its associated tribes. But what was the evidence? Babbitt hadn't cited any evidence. Owsley and the other plaintiffs had assured Barran that it is virtually impossible to establish cultural affiliation between Kennewick Man and any modern population, much less the Indian tribes that had claimed him.

By couching his decision in language that is required by the statute, like "preponderance of the evidence," Babbitt had made the decision look and sound correct. On its face, the decision paid lip service, she thought, to the wording required under the law. But the evidence to support the wording had been omitted.

Babbitt's letter concluded, "This determination of disposition to the claimant Indian tribes under NAGPRA precludes any study of the remains by the public. The claimants have been found to be the legal custodians of the remains and study may only be conducted with their permission."

Barran put down the letter, knowing that the newspaper reporters awaited her return calls to comment on the story. Babbitt's ruling would grab front-page headlines in the morning paper. She knew the spin would be that the scientists had lost. Yet they hadn't. Barran wanted an immediate counterattack to balance the spin. She figured the best way was to get something filed in court that afternoon, forcing the press to reflect that in the headline. She looked at her watch. It was 1:30. Time was not her ally. The courthouse closed in three hours.

Rather than calling back the press, Barran jotted down a list of legal issues raised by Babbitt's decision, ones that she could cite in order to request a status conference with the judge. While she produced a list, Schneider telephoned Cleone Hawkinson and told her to send an urgent E-mail to all the plaintiffs, notifying them of the news. As he hung up with Hawkinson at 1:50, Schneider received a fax from Barran, listing six questions raised by Babbitt's decision letter. Schneider quickly reviewed the list, made some edits, and faxed it back to her. Barran typed up a final version and passed it to her secretary to put on the firm's letterhead. "Get a runner ready to go to court," Barran told her. "This has to be filed before they close the clerk's office."

Then Barran called Schneider.

Reading from her computer screen, she rattled off the things she intended to ask from the court, concluding with an opportunity to challenge in court Babbitt's ruling.

At 4:30, an assistant in Barran's law office filed the request at the U.S. District Court in Portland.

The following morning

As Barran picked up the *Oregonian*, she immediately spotted the headline on page one: KENNEWICK MAN BELONGS TO FIVE TRIBES, U.S. SAYS. She quickly read the first two paragraphs of the article, then turned the page, spotting a new headline: SCIENTISTS' LAWYERS FILED A CONFERENCE REQUEST WITH THE COURT. Her strategy had worked; the government did not dominate the headline. "After four years of the federal government ducking and dodging this issue, we'll now have our day in court to challenge their decision," the story quoted Schneider saying.

Beneath Schneider's quote, Barran spotted a quote from Umatilla tribal leader Armand Minthorn. "This is a truly great victory for us," Minthorn said. "It gives me a tremendous feeling knowing that this Ancient One has been reaffirmed as one of our ancestors. As tribal people who have lived on the Columbia Plateau for thousands of years, we are eager to rebury our ancestor and give him back to the Earth."

Schneider sent the story to Owsley at the Smithsonian. Owsley didn't mind Minthorn's comments. Minthorn was claiming that his religion told him Kennewick Man was Indian. People, Owsley felt, should be free to believe what they want to believe. But no one, Minthorn included, has a right to force his beliefs on others, particularly religious beliefs. To Owsley, however, this dispute was not really over religion; it was over politics. Owsley had always figured that science would take a backseat to politics as long as Kennewick Man's future was in the federal government's hands. With Babbitt's decision issued, Kennewick Man's destiny finally shifted to the hands of Judge Jelderks. Law, Owsley hoped, would be friendlier to science than politics had been.

He asked Schneider when a court date would be set. Months down the road, Schneider predicted.

35

SHOW TIME

_____

June 18, 2001
Portland, Oregon

Trailed by Barran, Owsley entered the spacious glass-enclosed conference room on the twenty-third floor in her law office. The other scientists—George Gill, Richard Jantz, Dennis Stanford, Gentry Steele, Loring Brace, and Rob Bonnichsen—were seated around a massive cherry-wood conference table in front of a window that offered a breathtaking view of Portland and Mount Hood. Beneath the window, cans of soda and bottled water sat in a bucket of ice. Alan Schneider stood at the head of the table, sipping a can of Sprite and grinning. For five years he had longed for the day when all the scientists—his plaintiffs—would get their day in court. They were less than twenty-four hours away.

In preparation, Schneider and Barran had asked the scientists to come in for a briefing on what to expect. Instead of their case going before a jury as it would in a trial, they were going before a judge in a hearing. The court's primary role was to review whether Bruce Babbitt had acted lawfully when he determined that Kennewick Man should be returned to the tribes. Lawyers from the U.S. Justice Department would appear on Babbitt's behalf and argue that he had complied with the law. Barran and Schneider planned to argue that NAGPRA did not apply in this case because Kennewick Man was not Native American.

After Owsley took a seat, Schneider and Barran joined each other at the head of the table. "On behalf of Paula and myself," Schneider began, "I'd like to thank all of you for showing up in what we hope will be a red-letter day for science in America."

Her black suit perfectly pressed, Barran surveyed her clients. In her legal career she had never had so many brilliant minds in one room. "Let me give you a little bit of an outline of what's going to happen tomorrow," she said.

Barran began with security measures. The high profile and controversial nature of the case promised to attract spectators, protestors, and a large media contingent. The courtroom had only 150 seats. Lines might form outside the courthouse early in the morning. Two armed federal marshals were assigned to the courtroom. Barran advised the scientists to arrive in a group. A courtroom deputy would escort them in. Photo ID was required. Tape recorders, personal computers, and cell phones were prohibited. They would be required to pass through a metal detector. Barran glanced at Dennis Stanford, seated to her right, his black jeans held up by suspenders with metal clips. "The suspenders will have to go," she said, smiling. "They will set off the detector."

Owsley jotted down everything Barran said.

One by one, she talked the scientists through the major issues that she and Schneider expected Jelderks to address. The scientists wanted to know what their chances were of prevailing. Schneider assured them that their case was strong. The law was on their side. But lawsuits are like gambling, unpredictable. It was impossible to predict what a judge might do.

The media, on the other hand, are very predictable. Barran warned the scientists that the courtroom would be packed with reporters, some of whom would approach them for comments. She was sure they would try to frame this case as the scientists versus the Indians. Dennis Stanford rolled his eyes. He, Gill, and Owsley had dedicated the majority of their careers to helping tribes recover and identify their dead and recovering their culture. He had no patience for the press's tendency to pit him and his colleagues against the Indians.

"You will get asked," Barran said, "'Why are you doing this?'"

The group debated how to respond. Schneider offered some suggestions. As the discussion continued, Owsley remained silent. He never worried about what to say to the press. His habit was simply to tell the truth and let the chips fall where they may.

But Owsley had another concern. He believed that he and his colleagues would prevail in court. That, he feared, would trigger a backlash against science. He raised his hand for permission to speak. Everyone stopped talking. "I think that our real problem is that there's going to be a tremendous lobbying effort to tighten the NAGPRA law," he said. "Many of our colleagues are all too willing to duck and dodge. They will continue to do that. I see the Native American lobby coming in with lots of lobbying bucks. My real concern is that we may really outdo them here, but our colleagues will remain silent and we'll lose the battle."

No one disagreed.

"I think that covers it," Schneider said.

Owsley raised his hand again. "We're all very excited," he said. "But however it turns out, I want you to know on behalf of all of us how much we appreciate what you've done for us." He started clapping and all the scientists joined in applause. "We could not have got this far on our own."

After everyone left the room, Owsley remained behind, staring silently out the window at Mount Hood, his eyes fixed on a point. Observing him alone in the room, Barran stepped back inside. She marveled at how he managed to stand out even among a group of rare scientists. She walked over and stood beside him, wondering what he was thinking. He was merely marveling at the view.

The next day

One by one, the scientists filed into a jury box to the right of the judge's chair. Barran and Schneider took their place at a table in front of their clients. As Schneider reviewed his notes, Barran read an E-mail she had just received from her brother, entitled "What a Difference 30 Years

Makes." It said: "1970: Growing Pot. 2000: Growing potbelly. 1970: Rolling Stones. 2000: Kidney Stones. 1970: Peace sign. 2000: Mercedes logo." She laughed to herself, then looked up as the Justice Department lawyers strolled into court. She immediately concluded that the E-mail was right on the money. The Justice Department had sent Generation X attorneys from Washington: young, cocky, and dressed to the nines. Chewing gum and sporting a blue checked suit with a hip blue shirt and colored glasses, lead attorney David Shuey was flanked by a suntanned assistant in a form-fitting skirt and pink spandex sweater. One of Shuey's colleagues, tribal attorney Rob Smith, had sideburns shaved to a lightning-rod point and a wool suit with a green tie and green shirt. Together, they looked like a GQ model and a poor imitation of Brad Pitt. Barran and Schneider, products of the sixties, felt old and conservative.

Suddenly, Judge Jelderks entered the courtroom through a nine-foot-high wooden door. "All rise," the clerk said, banging the gavel.

Owsley leaped to his feet and stood motionless, staring at the judge as if at attention.

Perched in a high-backed chair elevated above the courtroom, Jelderks then asked Barran to start by briefly summarizing the scientists' argument.

She stood up and rattled off the highlights of their case.

Jelderks then turned to the government's lawyers. David Shuey pushed aside his oversize paper coffee cup and pulled the microphone closer to his mouth, not bothering to stand up. "Basically our argument is that the plaintiffs have no right to study these human remains in the custody of the United States," he said. "In terms of NAGPRA determinations, we believe the Secretary's determination that these remains were Native American is fully supported by the record."

A confused look crossed Owsley's face. This was a landmark case, yet Shuey could not have been more nonchalant. Owsley wondered if he even cared about the skeleton or the outcome.

Jelderks asked if both sides agreed that Kennewick Man was found on federal land.

Both agreed.

"I took a look at the state of Oregon to see what portion of Oregon is federal land that would trigger issues like we have today if there were remains found at some point in the future," Jelderks said.

He turned to his clerk and asked him to unveil an oversize map of Oregon situated at the head of the courtroom. All the federal territory within the state was shaded. That encompassed more than half of the state's land. The illustration was clear. The odds were that any ancient skeletons found in the future in Oregon were more likely than not going to surface on federal land. And any remains found on federal land would trigger the federal NAGPRA law.

"So the threshold issue in this case," Jelderks said, "is the definition of 'Native American' as Congress used the term in NAGPRA." The answer to that question would impact all future discoveries of human remains on federal lands.

Jelderks had the clerk dim the courtroom lights and turn on an overhead projector. The definition of "Native American" from the NAGPRA law appeared on a large screen at the head of the courtroom. It read: "Native American means of, or relating to, a tribe, people, or culture that is indigenous to the United States."

The lawyers from the two sides disagreed over the interpretation of the definition. The government argued that any remains predating the arrival of Columbus were by definition indigenous and therefore Native American. Under this interpretation, it didn't matter how recently a contemporary tribe, such as the Umatilla, had moved into the geographic region where Kennewick Man surfaced, or whether Kennewick Man was culturally or biologically related to the Umatilla. As long as Kennewick Man predated any documented presence of Europeans in North America, he had to be Native American.

Barran and Schneider disagreed, insisting that Congress had not defined "Native American" by a calendar year, choosing instead to use the phrases "relating to" and "that is indigenous." These words, they argued, called for some proof of a relationship between discovered remains and an existing tribe.

Jelderks questioned whether Congress had intended the term "Native American" to require some proof of relationship between discovered ancient remains and a present-day American Indian tribe. The government said no connection was required. The scientists said a connection was essential. Jelderks asked Barran to support her argument.

She started with the plain wording of the law. If Congress had wanted to define Native American by using the 1492 date, it would have put it in the definition. But NAGPRA's definition of Native American is silent with respect to dates. Barran pointed out that the first mention of 1492 as a cutoff year for determining which skeletons are Native American arose in 1997, a year after the scientists had filed their lawsuit. By that time, Barran and Schneider had raised serious questions about the identity of Kennewick Man, and the Interior Department had asked National Park Service archaeologist Dr. Francis McManamon to aid them in the case. McManamon then wrote an advisory letter that said any remains older than 1492 were pre-Columbian and therefore Native American.

"The first thing Dr. McManamon did," Barran said, "was forget what Congress said and forgot that his obligation was to back what Congress wrote: 'Native American means of, or relating to, a tribe, people, or culture that is indigenous.'

"Dr. McManamon says we don't need that 'relating to' stuff."

Justice Department and tribal attorneys smirked and rolled their eyes as Barran criticized Dr. McManamon's 1492 rule. "If you find a Viking civilization in Maine," she said, "they will immediately be turned over under NAGPRA because it will predate 1492 because we know the Vikings were exploring that coast in the year 1000." Taken to its extreme, the 1492 rule would even require an obviously European skeleton that is twelve thousand years old to be defined as Native American.

"Where is the analysis?" she asked, before answering her own question. There was no analysis. The 1492 rule, she argued, was nothing more than a desperate attempt to justify the Army Corps' initial decision to award Kennewick Man to the tribes. When it became clear that the corps had erred, the Interior Department expanded the definition of Native American in the NAGPRA law in order to ensure that Kennewick Man fit within it. And that, she argued, was unlawful. The Supreme Court had ruled that an agency did not have authority to deviate from the words of Congress unless Congress's words are ambiguous.

To Barran, there was nothing ambiguous about the definition Congress had drafted for Native American. She repeated the definition: "'Native-American means of, or relating to, a tribe, people, or culture that *is* indigenous to the United States.'" The words were plain. The word *is* required proof that Kennewick Man *is* related to a presently existing tribe or people. If, on the other hand, he was related to a tribe that was at one time indigenous but no longer exists, Kennewick Man did not fit the definition of Native American under the statute and should not be returned to a tribe to which he shared no affiliation.

"I'm sure before the end of this morning I will be sounding like Clinton, depending on what 'is' is," she said.

Jelderks grinned.

The opposing lawyers did not.

Barran talked faster. "Forget Kennewick Man for a moment," she said. "Suppose we had a skeleton that radiocarbon-dated at exactly 1492." The scenario exposed the flaw in the 1492 rule. Radiocarbon dating is not precise, leaving in question whether a skeleton dated at 1492 was actually forty years younger or forty years older. "Why would we treat 1490 differently from 1492, except all of us went to primary school and learned a little ditty that 'In 1492, Columbus sailed the ocean blue'?"

Expressionless, Jelderks pressed his index finger across his lips.

"NAGPRA requires a relationship of some sort, and it requires a group that *is* indigenous," Barran said, relentlessly making her point. "Present tense indicative. That *is* indigenous. The agency can't turn around and say, 'We interpret that to mean 1492, even if there is no relationship, and even if there is no group that is indigenous that we can point to.'"

"I think we all would have to agree," Jelderks said, "that there is nothing in the statute that creates a presumption that remains of a certain age are presumed to be Native American."

Jelderks gave Barran a hypothetical: If science confirmed that a group had occupied land in the United States ten thousand years ago, then became extinct five thousand years ago—therefore satisfying the definition of "indigenous"—would a modern American Indian tribe have a claim to such remains?

"If we have a tribe that had died out entirely and that is extinct," she said, "why would Congress be interested in repatriating that set of remains to a people that have no connection with those remains? If it had died out and been extinct, it has no modern-day relatives, no modern-day progeny. We don't have a link between those two groups."

Barran sat down and Shuey stood up. He began by saying that Congress had intended NAGPRA to apply to all prehistoric remains. Owsley scribbled the word "prehistoric" on his notepad, waiting to hear what Shuey said next. But Jelderks cut him off.

"Let's stop on that," Jelderks said, "because words are important. 'Prehistorically,' as applied to this case, simply means prior to recorded history, correct?"

Owsley was eager to hear Shuey's response. Shuey hesitated. "It would be prior, really, to European exploration," Shuey said. "'Prehistoric' means before there was written history. Who wrote the history of this country? Europeans, not Native Americans."

George Gill lowered his glasses to the tip of his nose and glanced at Owsley. They had a broader view of the definition of prehistoric.

"Some people, when they hear the term 'prehistoric,' think of dinosaurs," Jelderks said.

"I think it is very important when you are talking about prehistoric, we define it by European culture because it is when the Europeans arrived here that we have a written history," said Shuey.

Barran and Schneider exchanged notes. Shuey was making their argument for them. By saying that any remains predating the arrival of Columbus were prehistoric and therefore Native American, Shuey had essentially confirmed their allegation: that the government relied on a 1492 cutoff date to decide which skeletons were Native American.

Yet Shuey kept denying the existence of a 1492 rule. He quoted from a Department of the Interior memorandum that had previously been supplied to the court. "DOI defines 'Native Americans' as those 'tribes, peoples, or cultures that were present in the United States prior to documented European exploration.'"

Shuey was trying to have it both ways. While insisting there was no 1492 rule, he was saying that anything prior to the arrival of Columbus had to be Native American. And he offered no explanation for how the Interior Department would treat Viking remains. Questions about Vikings, he insisted, were irrelevant in the Kennewick Man case. Shuey pointed out that the Viking explorations happened in Maine. "We are talking about remains that were found in Oregon," he said.

"Washington, actually," Jelderks said.

"In Washington. Excuse me," Shuey said, repeating his point about a Viking skeleton. "It would not be considered Native American because it would not be related to tribes, people, or cultures indigenous to the United States."

Jelderks stopped Shuey again. "Maybe 'indigenous' is a word that sounds simple, [but] isn't quite so simple." Jelderks offered an example. His yard, he said, contained a species of plant that people in the area insisted had been around since the dinosaurs. "Would we all agree that that plant was indigenous?"

"Correct," Shuey said.

"To Portland, Oregon?" Jelderks probed.

Shuey didn't answer.

Jelderks probed further. He asked Shuey to assume that other than the old plant species in his yard, all other plants and trees in the Portland area were the result of seeds that were transported to the area by birds, people, or the elements. Although those seeds didn't originate in Portland, at some point in time the plants and trees those seeds produced would be defined as indigenous to Oregon and the United States.

Jelderks gave Shuey a hypothetical to consider. A group of blond-haired, blue-eyed people is discovered frozen in an ice cave atop Mount Hood. They lived in the region for generations before getting caught in an ice storm. "Why wouldn't they be Native American?" Jelderks asked. "They lived here; they resided here."

"If your hypothetical includes that they were here for several generations, then I think they would be indigenous." Shuey said it would be necessary to look for other evidence in the ice cave, such as artifacts, to determine whether the people were just passing through the area from someplace else.

"But to say that based on the color of their hair or the color of their skin or the color of eyes that they are not Native American," Shuey said, "I don't think anyone can buy that."

As was the case with Kennewick Man, Shuey said there were a variety of types of evidence to consider when determining its origin. But morphology—the study of the structure of humans—was not among them. "The fact that the cranium of these remains is not the same cranium that you would find in a present-day tribe today is not determinative whether those remains are Native American," he said.

Owsley immediately picked up on Shuey's backhanded slight. Without mentioning Owsley or Jantz by name, Shuey had attacked the credibility of their scientific conclusion. They had compared Kennewick Man's skeletal structure to the skeletons of thousands of Native Americans and confirmed that he was dramatically different. As a lawyer, Shuey did not personally handle or study human skeletons. He had grossly misrepresented the importance of skeletal structures as a key piece of evidence when determining what human population a particular skeleton belongs to.

Schneider scribbled, "Does he sound totally snotty to you?" on a yellow sticky pad and slipped it to Barran. "Just like the nerd who knew it all in high school!" she wrote back.

Itching to speak, Owsley kept a straight face. Next, Shuey disputed the scientists' suggestion that Kennewick Man could have migrated to North America, by boat or some other means, and merely been traveling through the area at the time of his death. To prove it, Shuey referenced the lithic point found lodged in Kennewick Man's hip. Shuey insisted it was a Cascade point, indicating that it was associated with the geography of the Kennewick area. Additionally, Kennewick Man had a marine diet, which Shuey said was consistent with eating salmon, a fish that is known to travel the Columbia River, on the banks of which Kennewick Man had surfaced.

Owsley immediately identified the scientific flaws in Shuey's conclusions. But the errors were not what caught Owsley's attention. It was the fact that Shuey was making a scientific argument for why Kennewick Man was indigenous to the area. Yet he was simultaneously taking the legal position that under NAGPRA all prehistoric skeletons—which he defined as pre-Columbian—should be defined as Native American. In such instances skeletons would not be subject to any study. But without some study, no one would have known that Kennewick Man had a lithic point in his hip or a marine diet.

"Let me ask you this," Jelderks said, returning to his hypothetical of remains surfacing in an ice cave on Mount Hood. This time Jelderks changed the fact pattern. He wanted Shuey to assume that scientific testing had confirmed that the remains in the cave were four hundred years old. "Do we accept as absolute fact without analysis that those 400-year-old remains would be related to modern American Indians rather than some other culture from some other part of the world?"

"You mean based solely on age without any other evidence?" Shuey asked.

"Let's make them 600 years old," Jelderks said.

"They are found in a vacuum with no sort of other evidence?" Shuey asked.

"Right."

"Were they found here on Mt. Hood?"

"Mt. Hood," Jelderks said.

"I would think that there is not any European exploration documented in this area until long after 600 years ago. If that were the case, I think the presumption would . . . be Native American, in the absence of any other evidence that suggested they were not."

"What would that evidence be?" Jelderks asked.

Shuey used Kennewick Man as an example. When government scientists performed carbon testing on Kennewick Man's bones, they found evidence of a marine diet.

Jelderks wanted Shuey to stick to the six-hundred-year-old remains in his hypothetical. "Since those remains are . . . clearly prior to documented Europeans in the area, regardless of what's with them or what their physical characteristics are, those would be Native American by definition," Jelderks said. "I don't understand what the criteria would be to apply other than age to show that those are not Native American unless we require some relationship to modern American Indians."

Shuey concluded that age would be determinative.

Now it was clear. Under the federal government's interpretation of NAGPRA, any future skeletons found on federal land and predating 1492 would be treated as Native American.

Jelderks asked Shuey to consider one more hypothetical involving the group in the ice cave: that before the group froze to death, its members had lived in the area long enough to be considered indigenous. However, every member of the group died in the cave, leaving behind no progeny. As a result, the group was extinct and had no affiliation to modern Native Americans. "Would that group be considered Native American or not?"

"Yes," Shuey said, explaining that the remains would be subject to NAGPRA.

"Those people then would never be studied unless the particular tribe or coalition of tribes . . . agree to that," Jelderks said.

"Congress gave that determination to the tribes," Shuey said.

In other words, scientists, archaeologists, and historians exploring the question of whether more than one ancient migration took place in the Americas would be precluded from looking at ancient skeletons, the most primary source of evidence. Anything predating 1492 would forever be treated as Native American.

Owsley grinned. The flaws in Shuey's argument were glaring. And it went beyond the example of Vikings. Owsley had personally examined human remains of Portuguese fishermen who had reached Canada before Columbus reached the Americas. The Portuguese had migrated by boat. To Owsley, Shuey's view of the peopling of the Americas was quite shallow.

Jelderks asked Shuey for clarification. "As of today it is the position of the Department of Interior that if there was a prior group of people living in what's now the United States who are no longer around—became extinct, for whatever reason—and with no identifiable connections with modern Native Americans, that they would still be classified as Native Americans under NAGPRA."

"That's correct," Shuey said.

Jelderks looked at his watch. It was after 12:00. He had one last question before recessing for lunch. "As you noticed, Ms. Barran did make significant issue of 'is' and 'was,'" Jelderks said. "If the statute said 'is or was indigenous to the United States,' it would be much easier for me to accept the Department's position on this. How can I accept the DOI's definition without, in effect, inserting a couple more words in the statute?"

He asked Shuey to consider the question over lunch.

Witnessing the justice system in action, Owsley was not sure he liked what he saw. He had spent his entire career at the Smithsonian studying skeletons. His groundbreaking work had opened a new window onto America's ancient past, shedding light on the possibility of multiple migrations to North America. The Justice Department had dispatched lawyers to try to stop future study, casting doubt over the ability of future scientists to flesh out the answers. The future of science in terms of ancient skeletons hung on the definition of the word *is*.

During the lunch recess, Barran and Schneider went back and read Bruce Babbitt's formal decision awarding Kennewick Man to the tribes. In it they found words that flatly contradicted what Shuey had just told the judge.

Following the break, Shuey stood at the microphone.

"I don't think that there is a difference between 'is' and 'was' in this instance," Shuey said, arguing that something indigenous, even if it becomes extinct, is still indigenous.

Jelderks stared at Shuey. He had had the entire lunch recess to come up with a better answer.

"Okay," Jelderks said, breaking the long, awkward silence.

Barran couldn't wait to rebut what Shuey had said before lunch. She had observed that the hypotheticals put to him by Jelderks had a common theme: surprises or unexpected discoveries that challenge traditional thinking.

She stepped to the microphone and pointed out that her clients were in the business of uncovering surprises. The fields of archaeology and anthropology were driven to discover. Barran told Jelderks about Owsley's recent experience at Jamestown. The revelation that Owsley had identified Africans captivated the courtroom. None of this had yet been published in American history books. Barran explained that the skeletons had initially been presumed Native American and had been on the verge of being repatriated. Jelderks leaned forward in his chair.

"I raise this to point out the fact that throughout this, much of the debate has been characterized as science for the sake of science," she said. In the Jamestown case, scientific inquiry prevented a rare window into African American history from being forever closed. Kennewick Man, she suggested, was a lot like Jamestown. It was a surprise discovery that challenged the traditional view of history. A skeleton this rare was too important to cavalierly turn over to Indian tribes without sufficient proof that he was indeed a Native American. She reached for Bruce Babbitt's ruling. "I think Your Honor should be rather shocked to hear Mr. Shuey's argument that there is no 1492 rule," she said, holding the paper. "This is from the Secretary himself. 'A series of radiocarbon dates clearly indicates a pre-Columbian date for the remains. It is reasonable to conclude that they are Native American.'"

The government's own documents confirmed that it had decided Kennewick Man was Native American based on its age. Scientific inquiry was not required.

Owsley shifted his eyes from Barran to the judge. He sensed that Barran was gaining the upper hand. Some of Shuey's colleagues appeared to be thinking the same thing. One of them passed Shuey a note. As he read it, Barran attacked Shuey's argument that the spear point in Kennewick Man's side and the marine diet were evidence that he lived in the area that he died in. Barran had asked Owsley about the spear-point injury. He explained that the bone around the point had regenerated, indicating a healing process. Owsley suspected that the point had been embedded in Kennewick Man's hip for up to twenty years prior to his death. Rob Bonnichsen studied CAT scans of the point and concluded it likely did not come from stone in the Kennewick region. These factors made it clear that Kennewick Man could have been stabbed somewhere far away from the place he died.

Barran also disputed Shuey's argument that a marine diet of salmon proved Kennewick Man was indigenous to the Columbia River region.

"Kennewick Man's diet does not necessarily have to come from the Columbia River," she said. "There are, by the way, salmon in Japan where Kennewick Man may have come from."

Shuey smirked.

Barran's point was clear. Definitive scientific research by experts was required. Shuey countered that the government had already conducted three forms of scientific testing: DNA, morphological, and taphonomic.

"But the Secretary didn't find anything in any of those three types of testing that would support his final conclusion, did he?" Jelderks asked.

"Well," said Shuey, "not in the DNA testing and not in the morphological or taphonomic testing."

Jelderks asked Barran her view on the testing done by government scientists. "I take it one step further," Barran said. "What came from the testing was evidence that Kennewick Man is unlike the claimants. In fact, his closest relations are Polynesian and Ainu of Japan."

Jelderks turned back to Shuey with a question. If scientists working for the U.S. government could not link Kennewick Man to the tribes, why and how did Babbitt? Shuey cited Babbitt's report. It said the evidence "reasonably leads to a conclusion of cultural affiliation."

"That was his final conclusion," Jelderks said. "But what specific pieces of evidence did he point [to] to support that conclusion?"

Linguistics, Shuey said. An expert linguist hired by the government said he could trace modern tribes' language back 4,500 to 5,000 years ago. "He believed that it was reasonable, and there was a strong possibility that he could take that language family back an additional 4,000 years to approximately 8,000 to 9,000 years ago."

"With no evidence, though, of what the language was 9,000 years ago?" Jelderks asked.

"That's correct," Shuey said.

The government had made some huge assumptions. First, it assumed that Kennewick Man was residing in—not passing through—the area in which he died. Second, it assumed that the language he spoke—whatever that was—had remained unchanged for roughly 9,000 years.

Jelderks detected another flaw. The NAGPRA law requires proof that a "shared group identity" exists between a present-day tribe and an "identifiable earlier group" to which human remains belong. Jelderks asked Shuey what the name of Kennewick Man's group was.

Shuey didn't have an answer.

Jelderks asked about the size of the group.

Again, no answer.

"Can we say what language he spoke?" Jelderks asked.

"I don't think we can say definitively."

The name, size, and language of the group to which Kennewick Man belonged were mysteries to Shuey and the Army Corps.

"There might well have been some evidence in the area of artifacts that would support a group culture," Jelderks said. "The Corps decided to put 500 tons or so of boulders and debris over the site. So that evidence is not available for either the Secretary or me, if it ever was available."

Jelderks checked the time. It was nearly 5 P.M. He was barely done with the definitions portion of the hearing. He asked both sides to reconvene in the morning. Shuey pointed out that the hearing was supposed to last one day. He and other Justice Department lawyers had morning flights back to Washington. Jelderks asked them to reschedule them.

36

WHITE HOUSE INVOLVEMENT

The next morning, the Justice Department legal team arrived in court early. While waiting for Jelderks, they stood at the table discussing their wardrobes and where they had gone drinking the night before. "How late were you at a bar last night?" Shuey asked one colleague, unaware that the microphone at his table was on. Barran and Schneider, who were seated at their table discussing strategy, glanced in his direction. The government lawyers fumbled to turn off the microphone.

Moments later Jelderks entered court carrying a stack of papers under his arm. He picked up right where he had left off, asking Shuey for evidence that linked Kennewick Man to the five tribes. Shuey had just one more suggestion: the tribes' oral traditions. Jelderks said he might accept oral traditions if Shuey could provide some specific stories or accounts that talked about Kennewick Man. Shuey had none.

Ready to move on, Jelderks turned to Barran. The most serious allegations raised in the scientists' brief accused the White House of applying political pressure on the Interior Department to ensure that Kennewick Man went to the tribes. Jelderks asked Barran to explain. She began with some history.

On August 30, 1996, the Army Corps had a telephone conversation with Umatilla leader Armand Minthorn. Barran read from an E-mail that a corps official wrote after the meeting. "I told him we will do what the tribes decide to do with the remains," the corps official had written.

The E-mail had been sent before anybody knew much about Kennewick Man and before the tribes had filed a claim. Then the tribes filed a claim, and the corps promised to return Kennewick Man to the tribes without having conducted any testing to determine his identity. But the lawsuit thwarted the plan. Once it was filed, Judge Jelderks directed the government to conduct a full, fair, and complete evaluation of Kennewick Man. This instruction put the government in an awkward political position, since the government had already promised Kennewick Man to the tribes. As a result, the government had tried to engineer evidence that would justify their promise while simultaneously burying any evidence that indicated Kennewick Man was not Native American.

A couple of tribal lawyers seated near Shuey laughed at Barran's charges, drawing a glance from Jelderks. But Barran cited a July 14, 1998, meeting between Department of the Interior officials and Umatilla leader Armand Minthorn. At it, they discussed Jelderks's court-ordered testing of the skeleton. The tribes opposed the testing, but the government had no choice but to carry it out. "Mr. Minthorn asked, 'If we get our answer on the first try, will we go forward with additional testing?'" Barran read. "The Department of Interior's response was, 'We will add language that if we get the right answer the first time, we will not go forward.'"

The other lawyers stopped laughing.

Barran reminded Jelderks that he had directed the government to conduct testing that would produce an unbiased, fresh look at Kennewick Man's origin. But the government, she insisted, had approached the testing with another goal in mind: getting the right answer for the tribes. Yet the government's scientists did not produce the result the government wanted. They concluded that Kennewick Man did not share a connection to any population groups currently residing on the North American continent. On the contrary, the testing indicated Kennewick Man most strongly showed links to Polynesia and Japan, a conclusion Owsley and Jantz had reached four years earlier.

Government and tribal lawyers fixed their eyes on Barran. One Army Corps official seated behind Shuey arched his neck upward.

"I think," Barran said, her voice slowing down to a hardly audible whisper, "you have a very clear record before you that every single ancient skeleton is irreplaceable, and this one is potentially magical because of its completeness. We don't have anything like it."

Every eye in the courtroom was riveted to Barran. Jelderks asked how many ancient skeletons from the Kennewick Man era were or had been available in the United States.

Barran looked Owsley's way and suggested to Jelderks that he be given a chance to speak. The last thing the government wanted to see was Doug Owsley testifying about ancient skeletons that he had studied, few of which were Native American.

"Your Honor," Shuey interjected.

Jelderks repeated his question to Barran. How many skeletons from the Kennewick Man era had been found in North America?

"Under a dozen," she said.

Jelderks was aware of only two, the Spirit Cave mummy and the partial skeleton from Idaho called Buhl Woman. He wanted more details about them. Barran deferred to Schneider.

Schneider explained that there were very few additional intact skeletons of the same vintage as Kennewick Man. None were as complete as Kennewick Man and the Spirit Cave mummy.

With so much evidence that Kennewick Man was not Native American, and with its being such a natural treasure, the government's decision to hand it over to the tribes raised the question: why? Where was the pressure coming from?

Barran had an answer: the White House.

Shuey rolled his eyes.

"We know the White House was involved in this case," she told Jelderks, causing scowls to arise on the faces of government lawyers. She said that at least six meetings took place between unnamed White House officials and Justice Department attorneys representing the Army Corps in the lawsuit against the scientists. Because there were few records and notes from those meetings, little was known about what transpired there. But Schneider and Barran were able to document the White House's influence in the decision to bury the discovery site under tons of rock. "We do know," Barran said, "it was the White House that put the covering up of this site on the fast track."

She had Jelderks's attention. He squinted his eyes as she referred to "secret meetings and secret talking points." Barran argued that the White House influenced the repatriation decision and that Interior Secretary Bruce Babbitt based his decision on political reasons rather than the scientific evidence.

Shuey dismissed this. He admitted that meetings took place. But he downplayed their significance. "In terms of contacts with the White House, it is a rather ambiguous term," he said. "I don't want anyone to get the impression that anyone at a high level was involved in these discussions."

"Let me back you up for just one moment," Jelderks said, reaching for a document that was part of the court record. He read it aloud. "'Colonel Curtis has been directed by the White House to get the bank stabilized. He wants the bank stabilization to be completed by placing rift-raft and finished no later than 1 January 1998.'"

Jelderks put the paper down. It appeared clear that the directive to bury the discovery site came from someone higher up at the White House than a visiting scientist. "Just knowing the way the government works, knowing how the military works," Jelderks said, "a colonel in the Army has been directed by the White House to get something done, the chain of command would indicate that that get done."

Shuey said he was not prepared to respond to the Colonel Curtis memo.

Jelderks turned back to Barran. He wanted more specifics. Who at the White House, or what agencies at least, were involved in secret meetings?

"We don't know because notes are so sketchy, and there are no notes from at least four of those meetings. So I don't know how counsel for the government can say nobody from a high level was involved because there isn't anything to tell us who was involved."

Jelderks grimaced as he listened, then called a brief recess after Barran concluded. Owsley walked out to the lobby and stared out a large glass window at the street sixteen stories below. The government and tribal lawyers huddled in a circle behind him, joking and laughing. Arrogance, he thought, supreme arrogance.

Suddenly Jantz walked up beside him. His plane was scheduled to leave in an hour, forcing him to miss the closing arguments. "I'm leaving," he said. "Carry on, Doug. Keep up the fight."

Most of the other scientists had caught flights earlier in the day. Besides Owsley, only Dennis Stanford remained. Owsley felt very alone. But he was used to that. He walked back into court and took a seat in the empty jury box. Five years earlier he had decided on his own to launch the fight to save Kennewick Man. That single decision had grown into an epic legal battle that pitted him against the combined forces of the Justice Department, the Interior Department, the Army Corps of Engineers, five Indian tribes, and the White House. His relentlessness had caused government officials to violate their own rules and policies. A national press corps had shined a spotlight on the case. Most important, Kennewick Man remained aboveground. One person, he had been taught as a boy, can make a difference.

In closing arguments, one tribal attorney summed up the sentiments of the rest. "Your Honor," he said, "there is simply no right to study under NAGPRA."

Shuey agreed. "Plaintiffs' claim seems to be that there may be some break in the historic record between these tribes and the remains that were found at the Kennewick site," he said, thumping the table and pointing his finger. "Plaintiffs cannot accept what the Indians know; that is, that their ancestors have always been there."

The pace of Shuey's speech escalated with each word. "We seem unwilling to believe that the Indians are the descendants of the people who were always here and want to hypothesize that there may have been someone else here, someone else who looks like us."

Struggling to keep pace with Shuey's speech, the court reporter checked his count. Shuey was talking at a clip of 280 words per minute, 80 words per minute faster than the average person speaks.

"Inadvertently or not, plaintiffs' theories are just another way to savage Indian heritage."

Owsley stared at Shuey. He had never met the man. But Shuey had shifted from a legal argument to a direct personal attack on Owsley. Yet Owsley was not permitted to defend himself.

But Barran could defend him. She was furious that Shuey had stooped to accusing Owsley and his colleagues of savaging Indians. After exposing the flaws in Shuey's statements, she turned her eyes from Jelderks to Owsley and Stanford in the jury box. "These are men who have devoted their lives to the study of human beings and people," she said. "They are passionate in their admiration for these cultures. It [their work] is not under any circumstances to be perceived or to be characterized by their own government as savagers of the heritage of this country."

Jelderks thanked all the parties and promised to issue his decision as quickly as possible. But he set no date. The case file had grown to over twenty-two thousand pages in five years. He wanted to read all of it before writing his opinion.

As the hearing recessed in Oregon, Benton County coroner Floyd Johnson was finishing up a law enforcement training seminar in Moses Lake, Washington. One of his colleagues there shared some disturbing news. The county building that houses the sheriff's department and the coroner's office in Kennewick was undergoing renovation. As a result, all materials from the evidence locker were put in temporary storage at an off-site location. Between the time that the evidence moved from the sheriff's office to the temporary site a mysterious box of bones turned up.

The news perplexed Johnson. He did not recall the presence of a shoebox in his evidence locker. Nor would it be easy to miss such an item in his locker, which measured approximately three feet in height, two feet in width, and four feet in depth. That night Johnson returned home from the conference. The following morning he tracked down the evidence technician and asked to see the newly discovered box. Inside he found femur bones. They were in what appeared to be the same Ziploc bags in which Jim Chatters packaged Kennewick Man's femurs nearly five years earlier.

The timing of the discovery was peculiar. Just two months earlier, on April 3, 2001, the FBI closed its investigation into the missing femurs after the U.S. attorney's office in Spokane, Washington, officially declined to prosecute the case for lack of evidence. Federal prosecutors had originally asked the FBI to investigate whether the bones had been stolen and if so, by whom. But besides Floyd Johnson, the FBI found few people willing to answer questions. Some members and employees of the Umatilla tribe who had access to Kennewick Man at the Battelle facility refused to be interviewed, or in some cases simply did not return repeated calls from investigators.

The FBI had a bigger problem. It was unable to document a chain of custody for the Kennewick Man. No one at the coroner's office, sheriff's department, Army Corps of Engineers, or the Battelle facility had conducted an inventory of the bones at the various points where custody of the Kennewick Man changed hands. Unable to confirm who had custody of the bones and when, the FBI could not determine when the bones disappeared, much less whether the disappearance was the result of a criminal act.

And now the bones had suddenly turned up.

Confused, Johnson called Chatters. The news dumbfounded Chatters. But he was even more surprised when Johnson described the box that contained the femurs. It was cardboard and labeled *Columbia Park II*. Chatters recognized the description at once. In the days following Kennewick Man's discovery, Chatters combed the discovery site and found other human bone fragments that were not part of the Kennewick Man skeleton. To avoid confusion, Chatters placed the non-Kennewick Man bones in a cardboard shoebox, labeled it *Columbia Park II*, and immediately turned it over to Johnson. It was weeks later that Kennewick Man, femurs included, was placed in a larger wooden box labeled *Columbia Park I* and turned over to Johnson. Both Chatters and Johnson agreed that the smaller *Columbia Park II* box was never put with or inside *Columbia Park I*. Yet when the *Columbia Park II* box turned up in Johnson's locker, it not only had the non-related Kennewick Man bones, but also contained the missing femurs from Kennewick Man.

For Chatters, something didn't add up. The night Johnson went to his home to take back Kennewick Man per the Army Corps's demand,

Chatters had personally placed the Kennewick Man femurs, along with the rest of the skeleton, in a large box with a screw-down lid. His friend Tom McClelland was his witness. Johnson was also present. Then they wrapped the box, labeled *Columbia Park I*, in evidence tape. Days later Johnson transported it to the Battelle facility. According to Johnson, the evidence tape was still wrapped around the box when he released it to Battelle, raising a question as to how the femurs got from one box to another.

There was another inconsistency. The second box surfacing in the evidence locker raised the presumption that it had always been there. But that seemed inconceivable. How could two large femur bones, each a foot long, that were valued at an estimated $25 million and the subject of a federal search have been sitting in Johnson's evidence locker all that time? Surely they would have been discovered. Yet the FBI had not searched Johnson's locker.

When the FBI learned of the discovery, it reopened the case. But the same problem remained, no paper trail. So in July 2001 the U.S. attorney's office again declined to prosecute the case for lack of evidence. Then on July 13, the FBI formally closed the case for good. No charges were filed. Nor was Jim Chatters, who was the last known person to handle the femurs, asked to examine the newly discovered femurs to confirm whether they were in fact the missing ones. Instead, the femurs were transported to the Burke Museum in Seattle and put with Kennewick Man.

37

WHAT YOU SEE HERE STAYS HERE

September 11, 2001
Jeffersonton, Virginia

As Owsley walked into the office off his living room, sunlight streamed in through the open windows. Birds chirped outside. There were no other sounds. He was not used to the quiet serenity of his newly purchased ten-acre farmhouse in rural Virginia.

At Susie's insistence, they had purchased the property—complete with horse stables, barns, fenced fields, and a vegetable garden— months earlier. After spending thirteen years in a cramped townhouse in the crowded D.C. suburbs, Susie decided they had to move to the country. With the children no longer living at home and Doug's extensive travel schedule, she was often home alone. She wanted horses and a farm to occupy her time.

Susie also figured the increased distance from the Smithsonian would keep Doug home more and help him develop some hobbies. He had gone his whole adult life without a hobby. He did not have time for one. The farm forced him to make time to tend the vegetable garden and to mow the acres of grass with a tractor, outdoor activities he thoroughly enjoyed. But even when he was weeding or riding the tractor, his mind was on bones. To him, an hour on the tractor was an hour to visualize a research project in his mind.

Nor did the new four-hour round-trip commute between the farm and the museum reduce Doug's time at the office. Instead, he revised his schedule, spending two days a week at home writing reports from his home office. The other three days he worked at the Smithsonian without returning home, choosing instead to sleep in his office to avoid the commute and maximize his work productivity. At nighttime he would unroll a foam pad that he kept tucked behind his desk and sleep on the floor.

Enjoying a day at home, he sat down at his desk. Suddenly the phone rang. He could see from caller ID that it was coming from his office at the Smithsonian.

"Doug," said his assistant Kari Sandness, "do you know what's going on? Turn on your TV."

He hung up and turned on the television. The Pentagon was on fire. A news broadcaster explained that a jet airliner had been purposely flown into the building.

The remote control in his hand, Owsley ran to the front door and yelled toward the barn for Susie. "You need to come inside right away!"

The urgency in his voice startled her. Doug never raised his voice. She left the horses and ran toward the house.

Inside she found Doug standing in front of the big-screen television. "What is it, Doug?"

"The Pentagon has been hit," he said, his eyes fixed on the screen. "We're under attack."

She looked at the screen. "Oh my God!" she said.

They both stood motionless. Then panic set in. "Where's Hilary?" Susie blurted out.

Hilary had recently started working as a budget analyst for the U.S. Navy. Her office was at the Pentagon. But she was sometimes assigned to off-site locations.

"I don't think she's working at the Pentagon right now," Doug said.

"No, Doug, I think she is."

They looked at each other. It dawned on them that they had no idea which part of the Pentagon Hilary worked in.

"Try her cell phone," Doug said.

Susie dialed it. She could not get a signal.

Suddenly the television coverage shifted to New York City. One of the World Trade Center towers had collapsed. Anger overcame Doug. The country, he thought, was at war. Someone had deliberately attacked American institutions. Hilary worked at one of them, and he could not locate her.

Their phone rang. Susie snatched up the receiver. "Owsleys."

It was their other daughter, Kim. She wanted to know if Hilary was safe.

Moments later the phone rang again. This time it was Susie's sister in Wyoming, calling to see if Doug was trapped in the city. Susie started explaining that Hilary was the one in the city.

"Get off the phone!" Doug said anxiously. "You've got to get off that phone so Hilary can call."

For the next two hours they sat on their leather sofa watching television. The longer they waited, the more they feared the phone call. Finally it came. The voice on the other end belonged to a complete stranger. "Who is this?" the man asked.

"Susan Owsley."

Doug moved closer.

"I'm calling for Hilary Owsley," the man said. "She wants me to let you know she can't get on the phone right now. But she's OK. She got out."

Hilary's office had been destroyed by fire. She and her coworkers had escaped moments before the ceiling collapsed.

Tears trickled down Susie's face.

The man hung up before she could get his name.

Doug wrapped his arms around Susie. His eyes welled up. Silent, they cried as television footage of New York City and Washington in chaos played on behind them.

Three days later

Doug's home office phone rang. It was Joe DiZinno from the FBI.

"Doug, this is the phone call you have been expecting."

DiZinno informed Doug he was to report to the Dover Air Force

Base in Delaware the following morning. The Armed Forces Medical Examiner's Office had activated a mortuary facility to receive all human remains from the Pentagon. Owsley's mission was a familiar one: separate and sort commingled remains, help select appropriate bone and soft-tissue samples for DNA typing and identification, and establish a personal identification and cause of death through analysis of skeletal pathology. He would work alongside a team of radiologists, dentists, and medical examiners.

As soon as he hung up, Owsley packed his duffel bag. The next morning he said good-bye to Susie, unsure when he would return.

When he arrived at Dover, he was told that he must abide by the Dover Code: What you see here stays here.

He donned rubber gloves, a mask, and a medical suit and approached the tables. The bodies were fresh. Decomposition was advanced. Sadness struck him like never before. He started to think about what he was seeing, the tragic finality that had come to hundreds of innocent people.

Without moving his lips, he talked to himself. Block it out, he thought. He had done this drill so many times before. Families needed him. They needed to have their loved ones identified. But his daughter's remains could have easily been in one of the body bags. At that moment, his job was the worst one in the world. But no one could do it better. He reached for a body. As soon as he touched bone, he hit his rhythm.

For the next seven days he worked twelve-hour shifts sorting and identifying the dead. Of the 189 Pentagon casualties, he worked on 60 of them.

When it was over, he did something he had never done. He wrote down his thoughts.

> The experience of Dover is part of the applied extension of my professional training, but it is on a different scale. It parallels my work at Waco on the Branch Davidians, or the help I provided in identifying soldiers killed in Operation Desert Storm, or the field and morgue work that I have done in Croatia on civilian casualties and mass graves from the Serbian invasion.
>
> From my perspective, the Dover work is truly hard core forensics.

We carefully evaluated remains that were frequently incomplete, often fragmented from having been torn apart by the impact, sometimes commingled with the tissues of others, and in some cases severely burned.

I numb out many of the harsh realities that I deal with in forensic investigations. But I will always remember how fortunate my family was not to have a child among those lost.

I will remember the faces as I saw them, as contrasted with the smiling faces seen in family photographs recovered as personal effects from wallets.

And I will always remember mortuary affairs staff patiently cleaning salvageable rings and jewelry so that they can be returned to families, trying diligently to remove all traces of the horrific event that caused the loss of a loved one.

When he finished, he gave the document to Hilary. He wanted her to have a record of his thoughts, something she could look back on years after he passed away.

A few months later, the Department of Defense awarded Owsley a medal and the Commander's Award for Civilian Service. It was the first time he had ever received a medal for his service.

38

IN DEMAND

April 2002

Nearly a year had passed since Judge Jelderks heard oral arguments in the Kennewick Man case. The twenty-two-thousand-page administrative record, and the judge's promise to study every page, had bogged him down. Barran and Schneider were itching for a decision. So was Owsley. But he had plenty to keep him busy in the interim. His star had risen. His services were in demand from agencies and organizations all over the country.

The National Park Service had sent him a molar that had been discovered on the Manassas National Battlefield Park, the scene of the First and Second Bull Run, two of the most famous Civil War battles. A visitor who was hiking along a trail at the park in May 2001 spotted a human jawbone fragment indented in soil, thinly buried. A park ranger filed a report with the Interior Department. Naturally, the Interior Department presumed the remains were of a Civil War soldier, a discovery that might merit a historical marker at the park.

The jaw fragment contained a molar, which had been sent by the Interior Department to the Armed Forces Institute of Pathology for examination. The Park Service sent it to Owsley for a second opinion. Owsley had exhumed or studied more Civil War skeletons than any scientist in the country—more than 150—from Gettysburg, Antietam, Port Hudson, and a host of other battlefields.

269

As soon as he saw the molar from the Manassas battlefield he knew it was not from a Civil War soldier. It was too recent. The tooth had a metallic filling. Although it was not unheard of for teeth to have fillings in the 1860s, the type of filling in this tooth was not from the Civil War era. It was too advanced. And the preservation in the tooth and the bone was too good. Owsley had seen enough fillings to know this was a post–World War II filling. But it was no more recent than the 1980s. He estimated the tooth was evidence of a missing person, someone who disappeared between 1940 and 1980. How the remains ended up at Manassas was another question, one that might require further excavation at the park.

Besides owing a report to the Park Service, Owsley had a report due to a Maryland state archaeologist on a case involving one of the nation's first Supreme Court justices, Gabriel Duvall. A delegate to the 1787 Constitutional Convention in Philadelphia, Duvall was chosen by President James Madison in 1811 to serve on the Supreme Court. Duvall died in 1834 and was buried in Maryland. Over time the graveyard, because of neglect and overcrowding, had become overgrown with brush, and rundown. In the 1980s some of Justice Duvall's descendants decided to have his body exhumed and reburied at another gravesite. To assist them, the family hired an archaeologist, who found an elaborate coffin in the area where the judge had been laid to rest. The skeleton was identified as Justice Duvall's and reburied with an appropriate land marker at a Maryland historical site.

A Maryland archaeologist called for Owsley's assistance when the Duvall family subsequently decided to relocate other deceased ancestors from the overgrown cemetery to the new graveyard where Justice Duvall had been reburied. Owsley's role was to identify one of the family members' remains. To get up to speed on the case, Owsley reviewed the file from Justice Duvall's reburial. In it he found pictures of Justice Duvall's coffin and the family-retained archaeologist holding the skull of Justice Duvall.

Owsley immediately noticed a problem. The skull purported to be Justice Duvall's had a full set of healthy teeth. Yet Justice Duvall lived into his nineties. In the early 1800s, people who lived into their nineties simply did not die with a full set of healthy teeth. Owsley noticed another inconsistency. The coffin hardware was not of the early-1800s vintage. It was much more recent. The Duvall family had exhumed and reburied the wrong guy. Owsley promised the state archaeologist a report for the Duvall family.

Besides a backlog of cases to report on, Owsley had something more burdensome pressing on him. His work on the Pentagon victims from 9/11 caused him to contemplate his own mortality. Despite spending a lifetime working at death scenes, exhuming graves, and handling lifeless human remains, he had rarely considered his own death. Now he did.

Raised Episcopalian and serving as an altar boy, he had been taught as a child to believe in God and the afterlife. Long before he became a scientist, however, he had stopped believing in both.

But he kept his true feelings to himself. He felt that religion served a critical purpose in American society, from teaching children moral values to helping solidify strong families. Owsley did not want to contribute to what he perceived to be a national trend, particularly among some scholars, toward diminishing the importance of religion. He had another reason for withholding his personal views. He routinely worked with family members who were mourning the loss of loved ones. Religion or a belief in life after death helped people cope with loss, pain, and the fear of dying.

Death did not scare Owsley. The unpredictability of the timing, however, suddenly had him anxious. In one respect, he was at peace with his own mortality; his personal relationships with his wife and daughters were excellent. Nothing was wanting in terms of his family. But from a professional standpoint, he had not achieved his desires.

He had looked at more than ten thousand individual skeletons, probably more than anyone in the world. Yet his vast knowledge was locked up in his own mind. He wanted to write it all down. His greatest desire—to publish a written record of the human populations he had studied—had eluded him. Indians, in particular, represented the largest percentage of skeletons he had examined, more than five thousand in all. In addition to publishing a comprehensive volume on American Indians, he wanted to put them into perspective with the other skeletons surfacing in North America, such as Kennewick Man, the Spirit Cave mummy, and the Clovis people.

A sense of urgency overcame him. With tribes filing NAGPRA claims to recover and rebury skeletal collections all over the United States, time was of the essence. He decided to return to Tennessee, the scene of his graduate school studies. The university still had the seven hundred Arikara Indian skeletons from the Larson site, the ones that Owsley had studied as a student. The Arikara had made a claim to recover the collection. Before it went back, Owsley wanted to reexamine it. For his master's thesis he had analyzed each skeleton to determine its age and sex. That marked the first time he had ever assigned ages to a skeletal collection. With decades of experience now under his belt, he wanted to recheck each skeleton for proper age.

May 2002
University of Tennessee

The older Owsley got, the better his vision became. He wore glasses and had trouble seeing far distances. But when it came to bones, his eyes were sharper than ever. Over a three-week span he analyzed all seven hundred Arikara skeletons from the Larson site, a process that had taken him nine months as a graduate student. He readjusted the age on many of his original assessments. This time he also detected many nuances in the bones that he had missed before, such as projectile points embedded in them. He also found much more evidence of tuberculosis in the children and bone cancer in the adults.

To help enhance the mortality portrait he hoped to document of Plains Indians, Owsley set aside some of the better bone specimens to bring back to the Smithsonian for advanced photography.

During breaks between analyzing the Arikara remains, Owsley spent some time on the Tennessee campus with two of its premier zooarchaeologists—archaeologists who study animal bones. Owsley had brought with him a tooth that had been shipped to him at the

Smithsonian by a paleontology firm in California. A construction company that dug up some bones outside Los Angeles had hired the firm, whose paleontologists determined that the excavators had uncovered a seventeen-thousand-year-old saber-toothed tiger den. The den was fourteen feet below the surface and contained the bones of two adult cats and numerous juvenile cats, as well as bison bones and other animal bones presumed to be remains of the tigers' prey. The discovery of one tooth mystified paleontologists, however. It appeared more human than animal. The firm sent it to Owsley for his expert opinion.

Owsley analyzed it and saw why it would be tempting to view the tooth as a human's. Upon closer inspection, he detected notches on both sides of the tooth. Corresponding notches were not on the chewing surface, suggesting that something had been inserted between the individual teeth for the purpose of performing some kind of task. In other words, the tooth reflected the wear of a person who had been using his teeth like a third hand, like a fisherman who inserts line between his teeth while using his hands to tie a knot or bait a hook.

But looks can be deceiving, and Owsley was reluctant to conclude that the tooth was human. The presence of a human tooth in a seventeen-thousand-year-old tiger den was hard to fathom. Owsley wanted to make sure that there was not some extinct animal out there that might have had teeth closely resembling those of humans. He asked both Tennessee zooarchaeologists to examine it.

They compared it to teeth from scores of extinct animals from that time period. Nothing matched. They were stumped. Owsley was intrigued, but not convinced. He wanted to consult other zooarchaeologists to be sure there was not some extinct animal that he might not be aware of. They agreed with Owsley. The tooth was human.

From Tennessee, Owsley headed to South Carolina to help the U.S. Navy identify the Confederate sailors from the *H.L. Hunley*, the first submarine to sink a ship in battle. After ramming a torpedo that downed the USS *Housatonic* in Charleston Harbor on February 17, 1864, the *Hunley* never returned to port, and mystery had shrouded its fate for more than one hundred years. Then in 1995 a dive team discovered the ship off the coast of Charleston. The skeletal remains of eight crew members were inside. In 2001, with the vessel in dry dock, Owsley had gone into the sub and excavated the interior sediments. He saw something no one else saw. Instead of mere scattered bones deposited in sediment on the ship's floor, Owsley visualized the men dying in 1864. In his mind he could see them at their stations in the tin-can-like iron sub as water penetrated. He could see the men going from life to death to advanced decomposition, and then in the final stage their bones coming apart, sinking to the floor of the sub, and settling in the sediment.

With the bones all removed from the sub, the time had come for Owsley to return to Charleston and rebuild the eight men, preparatory to identifying them and performing a proper burial.

39

THE DECISION

August 30, 2002
Portland, Oregon

"It's coming."

That's all Alan Schneider had to say when he called Paula Barran midway through the morning on the Friday heading into Labor Day weekend. Since arguing their case before Judge Jelderks 436 days earlier, they had eagerly anticipated his decision. The moment had arrived. Earlier that morning Alan had received several calls from reporters. They said the decision would be released by day's end.

Barran tried to remain reserved. For six years she and Schneider had battled the federal government, which had used twenty-five government lawyers and close to fifty engineers, archaeologists, scientists, and other personnel, and spent an estimated $3 million of taxpayers' money to prevent the scientists from studying Kennewick Man. Schneider logged over 6,000 hours on the case. Barran and her partners put in an additional 2,000. Combined, Schneider and Barran had run up ✳ roughly $1.5 million in legal fees. But they hadn't billed their clients a penny. Lawyers conducting a cost-benefit analysis would judge this case a colossal loss from a business perspective. But for Schneider and Barran the case was never about money. The federal government's efforts to bar scientific study of Kennewick Man were, in their view, a political

275

injustice that begged for a legal remedy. They never regretted their decision to underwrite the cost of the litigation. They just wanted to win. And waiting for Judge Jelderk's ruling on their motion to dismiss the Interior Department's decision that Kennewick Man belonged with the tribes had been like waiting for the biggest jury verdict in their careers.

When ready, Jelderks would post his decision on the federal court's Electronic Case Filing (ECF) system. Then ECF would instantly E-mail it to the accounts of the attorneys representing all interested parties. Using a security code password, Barran and Schneider would access a link to the decision and download the text. Barran told Alan that she and her husband Richard Hunt had plans to spend the long holiday weekend in central Oregon at their vacation home in Bend. They were set to leave the law offices just after noon to get a jump on the traffic. If the decision had not been posted by then, she would take her laptop with her for the weekend.

Schneider assured Barran he would call her on the road as soon as any E-mail came from the court. He instructed his secretary to log on and check their E-mail every thirty minutes. Soon he had her checking every fifteen minutes, then every five. By mid-afternoon, the secretary was logging on every three minutes. Schneider felt like a kid on Christmas eve; waiting for the moment it seemed would never arrive.

By 1:30 P.M. Barran and Hunt were in their BMW SUV headed toward Central Oregon. With Richard at the wheel, Paula plugged her cell phone in to the cigarette lighter and attached her earpiece. They were not outside Portland before she called her secretary to see if the decision had come. It hadn't. "Call me the instant you hear anything," Barran told her.

Driving through mountainous passes that continually disrupted service to Paula's cell phone, Richard smiled wryly every time they hit a stretch of road where the cell phone regained its signal. Each time, Barran redialed her secretary to make sure she hadn't tried to call while the phone was out of range. The secretary had no news to report. Paula's stomach was in knots.

At 3 P.M. Pacific Standard Time Richard glanced at the dashboard's digital clock. Then he looked briefly at Paula. "Elvis has left the building, you know," he said.

She knew what he was getting at. If Jelderks truly planned to release the decision on the Friday before Labor Day, he would probably exit the courthouse by 3 P.M., leaving behind instructions for his clerk to post the decision at 4:30, right when the courthouse closed for the day.

Nonetheless, Paula called her secretary right up to 4:30. Still, no E-mails had come from the court.

Then at 4:36 Barran's cell phone rang.

She fumbled for her earpiece.

It rang again.

"Answer it. Answer it," Richard said.

She slammed the earpiece in her ear and reached for the phone. "Where's the button?" she shouted, frantically trying to answer before losing the call.

"This is Paula," she said.

"It's here!" her secretary said. "You want me to read it?"

The decision was seventy-six pages long. Paula told her to scan down to the last page and read the conclusion.

"For the reasons set out above," the secretary began, "Plaintiffs' motion for an order vacating Defendant's decision on remand (#416-1) . . ."

"C'mon. C'mon," Paula whispered.

". . . is GRANTED."

"Son of a bitch," Paula said slowly, turning toward Richard and putting her thumb up. "We won!" she shouted. "We won. We won."

Richard laughed.

The secretary kept reading.

Paula kept yelling.

"Let me listen," Richard yelled. "Let me listen."

"Plaintiffs shall submit," the secretary continued, "a proposed study protocol to the agency within forty-five days."

"Holy shit!" Paula said. "We got study rights. Yes. Yes. Yes."

By now the secretary was yelling so loudly that Richard could hear her through Paula's earpiece.

"We won," Paula shouted. "We won. We won."

Paula hung up and immediately called Alan.

He had already read the first fifteen pages of the decision. "It's beautiful reading," he told her. "It's a home run."

As Schneider went through each detail with Barran, she shouted and cheered. Richard could not stop laughing. But he had a more global question on his mind: Was their victory decisive? In other words, how strongly was the judge's reprimand of the federal government?

It was so strong that the decision read more like an indictment.

To begin with, Jelderks found that the Army Corps of Engineers had failed to act as fair and neutral decision-makers from the moment it took possession of Kennewick Man. Under the Administrative Procedure Act (APA), the federal law governing agency conduct in decision-making situations, the corps is obligated to be fair and impartial toward all affected parties. Jelderks concluded by censuring the corps for engaging in a litany of "secret" communications with the tribes. In addition to secretly providing documents and advance copies of reports, the corps carried out private meetings with the tribes, meetings that were aimed at producing a preconceived result. At the same time, the corps ignored or rejected repeated requests from the plaintiffs to see the same documents and reports, essentially foreclosing them from information needed to respond to agency decisions.

Such actions are unlawful under the APA.

Jelderks saved his strongest condemnation for Interior Secretary Bruce Babbitt, who ultimately took control of the case from the corps and subsequently determined that Kennewick Man was Native American and belonged with five tribes in the Northwest. Babbitt's decision was virtually equivalent to passing a law. Under the APA, a judge cannot substitute his judgment for that of an agency, or set an agency's decision aside on the basis that a court might have reached a different conclusion. Agencies are afforded a great deal of deference. To rescind Babbitt's decision, Jelderks had to conclude Babbitt acted arbitrarily, abused his discretion, or failed to act in accordance with law. This required proof that Babbitt failed, as required by the APA, to demonstrate a rational connection between the facts before him and the choice that he made.

Jelderks excoriated Babbitt for failing to do just that. The only facts Babbitt relied on to judge Kennewick Man a Native American were that the bones were pre-Columbian and found in America. If Jelderks put his judicial seal on Babbitt's decision, it would set a precedent that would require that any skeleton found in the United States older than 510 years to be automatically defined as Native American, even if it had no relationship to a present-day tribe. Babbitt, Jelderks decided, had stretched the definition of Native American far beyond the one Congress identified under NAGPRA. Worse still, he made the stretch arbitrarily, without authority from Congress or soliciting proper comment and input from interested parties, which is a violation of the APA.

Babbitt's decision had other legal problems. He had determined that Kennewick Man belonged to five confederated tribes. Yet NAGPRA explicitly authorizes repatriation of remains to a single tribe, the one with the closest cultural affiliation. Implicit in this requirement is proof of tribal affiliation. "The Secretary's analysis contradicts the plain language of the statute," Jelderks wrote, chiding him for failing to make a real effort to analyze the claims separately.

Jelderks determined he had no choice but to set aside Babbitt's decision. The question was, what next? Normally, under the APA, a judge in this situation would remand the case to the agency with instructions to reevaluate the decision. But Jelderks took the extraordinary step of removing the jurisdiction of Kennewick Man from the Interior Department. He said it would be futile to give the agency another chance to reevaluate the evidence. It had proven itself unable to act in an unbiased fashion. "There is no reason to believe," Jelderks wrote, "that another remand would yield a different approach or result."

The move was a harsh rebuke for a governmental agency that had played fast and loose with the facts and violated the due process protections in the APA.

For Schneider, the decision vindicated his clients. Paula could not wait to read the decision for herself. She and Alan were still on the phone when Richard pulled up to their vacation home in Bend. "Let me call you back from a land line," she said, racing inside with her laptop. Moments later she was on line checking her E-mail account.

She scanned down to an E-mail with a link to the decision. She clicked on the icon and printed all seventy-six pages.

"You're going to be useless tonight," Richard said, carrying in their bags.

Museum of Natural History
Washington, D.C.

The halls of the museum were closed for the day and most Smithsonian scientists were off for the long holiday weekend. Not Doug. It was after 8:30 P.M. and he was still in his office, seated at an observation table reviewing the butchered bones of a hitchhiker who had been picked up and murdered by drug dealers. Shivering on account of a malfunctioning air conditioner that had reduced his office to an icebox, Doug sipped a cup of hot beef vegetable soup as he tried to write up the forensic report on the cause of the hitchhiker's death.

It occurred to him that Judge Jelderks had imposed a deadline on himself to issue his decision before Labor Day. Doug called Alan Schneider's assistant Cleone Hawkinson to see if anyone had heard anything from the court. Hawkinson broke the news to Doug that the judge had ruled in his favor and that he and the other scientists would get the chance to study Kennewick Man. Her words had not yet sunk in when Owsley's other line started ringing. It was Susie calling from the hospital. She had a rare break and just wanted to see if he was still at his office.

Doug told her that he and the scientists had prevailed.

His tone was measured, but Susie could hear the satisfaction in his voice.

"Congratulations," she said.

It was 1 A.M. by the time Susie and Doug met up at home. Exhausted, they went to bed without much discussion of the decision. The following morning Doug telephoned his father in Wyoming to give him the news. By that time, it had started to sink in. Doug felt an overwhelming sense of affirmation. The sense of relief was similar to what he felt when he finally earned his Ph.D., only this was greater. The satisfaction was akin to being hired by the Smithsonian, only this was richer. His legal and political saga to save Kennewick Man was a life-time achievement. He had risked his job and his reputation by stepping into the teeth of a politically charged controversy. No one forced him to sue the federal government. He volunteered to do it. And he won.

The victory was monumental and had ramifications far beyond Kennewick Man. But for Doug the triumph, like his motivations, was internal. He never doubted that Kennewick Man was not Native American and should not be repatriated to the tribes. The only question was whether to risk his career and reputation by standing up to the powerful political forces working to bury the remains. He alone was uniquely positioned to fight the fight. If he hadn't, Kennewick Man would have gone underground six years earlier and all future discoveries of his vintage would be at risk of repatriation without study.

I did what was right, he thought, feeling as though he had passed the most significant personal test of his career. And that's all that matters.

Owsley's Virginia farmhouse phone rang all weekend. Reporters called to get his reaction. Colleagues and friends called to congratulate him. Everyone wanted to talk. But he did not answer the calls. Nor did he feel deserving of accolades.

Wearing beat-up workboots, blue jeans, and a baseball cap, Doug spent his weekend picking the season's last crop of vegetables and pushing a rotor tiller through his garden. Alone, he was at peace, his hands working the machine and his mind visualizing the study of Kennewick Man.

SOURCE NOTES

The primary sources for this book include interviews conducted by the author, the personal papers of Doug Owsley, legal documents from the Kennewick Man litigation (*Robson Bonnichsen et al. v. U.S. et al.* Civil Case No. 96-1481), and Dr. Owsley's forensic case files. Owsley's personal papers included copious handwritten notes, personal correspondence, and logbooks chronicling his travels. Owsley's forensic files included dental and medical records and X rays, correspondence from state and federal law enforcement agencies, transcripts, personal correspondence between Owsley and surviving family members, crime scene photographs, Owsley's field notes, and forensic reports from each of the cases featured in the book.

Attorneys Paula Barran and Alan Schneider also provided the author with access to the complete court file from the Kennewick Man litigation.

Secondary sources included scholarly articles, press reports, and personal correspondence to the author.

The author also relied on background material in the form of textbooks, treatises, reference books, laws, maps, and photographs.

AUTHOR'S INTERVIEWS

Dennis Apodaca, Portland, Oregon • Paula Barran, Portland, Oregon • Bill Bass, Knoxville, Tennessee • Sam Blake, Portland, Maine • Randy Blake, Washington, D.C. • Idy Bramlet, Lusk, Wyoming • Ryan Brown, Kennewick, Washington • Jim Chatters, Richland, Washington • Jenny Chatters, Richland, Washington • Chip Clark, Washington, D.C. • Donald Craib, Washington, D.C. • Amy Dansie, Carson City, Nevada • Dolores Davis, Scranton, Pennsylvania • Joe DiZinno, Washington, D.C. • Richard Donaldson,

Philadelphia, Pennsylvania • Robert Fri, Washington, D.C. • George Gill, Laramie, Wyoming • Lauryn Grant, Washington, D.C. • Danny Greathouse, Washington, D.C. • Edwin Harnden, Portland, Oregon • Cleone Hawkinson, Portland, Oregon • Dan Hester, Louisville, Colorado • Audi Huber, Pendleton, Oregon • Richard Hunt, Portland, Oregon • Richard Jantz, Knoxville, Tennessee • Floyd Johnson, Kennewick, Washington • Suzanne Johnson, Kennewick, Washington • Rebecca Kardash, Washington, D.C. • Bill Kelso, Jamestown, Virginia • Bill Lehey, Richland, Washington • Sharon Long, Carson City, Nevada • Julie Longenecker, Pendleton, Oregon • Mike Lyon, RWC, California • Tom McClelland, Richland, Washington • Andy Miller, Richland, Washington • Henry Miller, St. Mary's City, Maryland • Armand Minthorn, Umatilla Indian Reservation, Pendleton, Oregon • Bill Owsley, Laramie, Wyoming • Douglas Owsley, Jeffersonton, Virginia • Hilary Owsley, Jeffersonton, Virginia • Kim Owsley, Jeffersonton, Virginia • Susan Owsley, Jeffersonton, Virginia • Audrey Pfister, Lusk, Wyoming • Karen Rehm, Jamestown, Virignia • David Riggs, Jamestown, Virginia • Kari Sandness, Washington, D.C. • Lee Sappington, Idaho • Alan Schneider, Portland, Oregon • John Schultz, Richland, Washington • Diane Stallings, Yorktown, Virginia • Dennis Stanford, Washington, D.C. • Pam Stone, Amherst, Massachusetts • Jeff Van Pelt, Umatilla Indian Reservation, Pendleton, Oregon • John Verano, New Orleans, Louisiana • Roz Works, Carson City, Nevada

To ensure the highest level of accuracy, most interviews were tape-recorded and personally transcribed by the author. In many instances, passages in the book containing dialogue were presented to the participants to proofread for accuracy prior to publication.

SCHOLARLY ARTICLES AND PUBLISHED PAPERS

Bonnichsen, Robson, and Alan L. Schneider. "Battle of the Bones." *The Sciences*, July 2000.

Collins, Michael B. "Clovis Second: Time Is Running Out for an Old Paradigm." *Discovering Archeology*, February 2000.

Dixon, James E. "Coastal Navigators: The First Americans May Have Come by Water." *Discovering Archeology*, February 2000.

Frison, George C. "Progress and Challenges: Intriguing Questions Linger after 75 years of Answers." *Discovering Archeology*, February 2000.

Gill, George W. "Two Mummies from the Pitchfork Rock Shelter in Northwestern Wyoming." *Journal of the Plains Anthropological Society*, 1976.

Gill, George W., and Douglas W. Owsley. "Electron Microscopy of Parasite Remains on the Pitchfork Mummy and Possible Social Implications." *Journal of the Plains Anthropological Society*, 1985.

Grisbaum, Gretchen, and Douglas H. Ubelaker. *An Analysis of Forensic Anthropology Cases Submitted to the Smithsonian Institution by the Federal Bureau of Investigation from 1962 to 1994*. Washington, D.C.: Smithsonian Institution Press, 2001.

Hall, Don Alan. "A Database on Humanity's Past: Smithsonian Team Races the Clock with Repatriation." *Mammoth Trumpet*, January 1997.

Haynes, C. Vance. "New World Climate." *Discovering Archeology*, February 2000.

Hinds, V. Strode. "Reconstructing Charles Floyd." *We Proceed On*, February 2001.

Hofman, Jack L. "The Clovis Hunters." *Discovering Archeology*, February 2000.

Kelso, William, and Beverly Straube. "1996 Interim Report on the APVA Excavations at Jamestown, Virginia." The Association for the Preservation of Virginia Antiquities, 1997.

Mansisidor, Julie. "Getting to the Bare Bones." *University of Wyoming Alumni Magazine*, fall 1999.

Miller, Henry. "Mystery of the Lead Coffins." *American History*, October 1995.

Owsley, Douglas W. "Fluctuating Asymmetry as an Indicator of Developmental Insatiability in Cleft Lip and Cleft Palate." *American Journal of Physical Anthropology* 52, no. 2 (1980).

———. "Identification of the Fragmentary, Burned Remains of Two U.S. Journalists Seven Years after Their Disappearance in Guatemala." *Journal of Forensic Sciences*, November 1993.

———. "Techniques for Locating Burials, with Emphasis on the Probe." *Journal of Forensic Sciences*, September 1995.

Owsley, Douglas W., and David R. Hunt. "Clovis and Early Archaic Crania from the Anzick Site (24PA506), Park County, Montana." *Journal of the Plains Anthropological Society* 46 (May 2001).

Owsley, Douglas W., and Richard L. Jantz. "Archeological Politics and Public Interest in Paleoamerican Studies: Lessons from Gordon Creek Woman and Kennewick Man." *American Antiquity*, September 2001.

———. "Biography in the Bones: Skeletons Tell the Story of Ancient Lives and Peoples." *Discovering Archeology*, February 2000.

———. "Putting a Paleoamerican Campsite to Bed." *Mammoth Trumpet* 17, no. 3 (June 2002).

Owsley, Douglas W., Robert W. Mann, et al. "Positive Identification in a Case of Intentional Extreme Fragmentation." *Journal of Forensic Sciences*, July 1993.

Owsley, Douglas W., and Sarah B. Pelot. "Three Grams of Bone and Three Dental Fragments Aid Identification of a Homicide Victim." *Journal of Forensic Identification*, September 1995.

Owsley, Douglas W., D. H. Ubelaker, et al. "The Role of Forensic Anthropology in the Recovery and Analysis of Branch Davidian Compound Victims: Techniques of Analysis." *Journal of Forensic Sciences*, May 1995.

Prag, John. "Putting Flesh on Bone: Facial Reconstructions Give New Meaning to Old Skulls." *Discovering Archeology*, February 2000.

Stafford, Thomas W. "How Old Is It? The Powers and Pitfalls of Radiocarbon Dating." *Discovering Archeology*, February 2000.

Stanford, Dennis, and Bruce Bradley. "The Solutrean Solution: Did Some Ancient Americans Come from Europe?" *Discovering Archeology*, February 2000.

Steele, Gentry D. "The Skeletons' Tale: Old Skulls Are Painting a Complex Picture of American Origins." *Discovering Archeology*, February 2000.

Tankersley, Kenneth B. "The Puzzle of the First Americans." *Discovering Archeology*, February 2000.

Ubelaker, Douglas H. "A History of Smithsonian-FBI Collaboration in Forensic Anthropology, Especially in Regard to Facial Imagery." Paper presented at the Ninth Biennial Meeting of the International Association for Craniofacial Identification, FBI, Washington, D.C., July 24, 2000.

———. "Ales Hrdlicka's Role in the History of Forensic Anthropology." *Journal of Forensic Sciences*, July 1999.

———. "J. Lawrence Angel and the Development of Forensic Anthropology in the United States." In *A Life in Science: Papers in Honor of J. Lawrence Angel*, edited by Jane E. Buikstra, pp. 191–200. Kampsville, Illinois: Center for American Archeology, 1990.

———. "T. Dale Stewart's Perspective on His Career as a Forensic Anthropologist at the Smithsonian." *Journal of Forensic Sciences*, March 2000.

———. "The Influence of William M. Bass III on the Development of American Forensic Anthropology." *Journal of Forensic Sciences*, September 1995.

U.S. Department of the Interior, National Park Service. "Jamestown Island Revisited." Vol. 22, no. 1, 1999.

Waters, Michael R. "Proving Pre-Clovis." *Discovering Archeology*, February 2000.

PRESS REPORTS

Begley, Sharon, and Andrew Murr. "The First Americans." *Newsweek*, April 26, 1999.

Blake, Samuel. "What Else Did the C.I.A. Know?" *New York Times*, March 30, 1995.

Brown, David. "Nevada Mummy Caught in Debate over Tribal Remains." *Washington Post*, May 5, 1996.

Claiborne, William. "FBI Probes Theft of Ancient Bones Sought by Tribes." *Washington Post*, September 30, 2000.

Coll, Steve. "The Body in Question." *Washington Post Magazine*, June 3, 2001.

Dawson, Jim. "Bones Are Building Block toward Filling in Full Human." *Minneapolis Star Tribune*, March 17, 1993.

Egan, Timothy. "Expert Panel Recasts Origin of Fossil Man in Northwest." *New York Times*, October 16, 1999.

Gore, Rick. "People Like Us." *National Geographic*, July 2000.

Hill, Richard. "Kennewick Man Belongs to Five Tribes, U.S. Says." *Portland Oregonian*, September 26, 2000.

Lemonick, Michael. "Bones of Contention." *Newsweek*, October 14, 1996.

Malcomson, Scott L. "The Color of Bones: How a 9,000-Year-Old Skeleton Called Kennewick Man Sparked the Strangest Case of Racial Profiling." *New York Times Magazine*, April 2, 2000.

Manning, Anita. "Smithsonian Scientist Knows the Truth in His Bones." *USA Today*, June 24, 1997.

Milloy, Ross E. "Years after Cult Fire, Legal Battles Gather Force." *New York Times*, November 26, 1999.

Nemecek, Sasha. "Who Were the First Americans?" *Scientific American*, September 2000.

Parfit, Michael. "Dawn of Humans." *National Geographic*, December 2000.

Potter, Maximillian. "The Body Farm." GQ, June 2002.

Preston, Douglas. "The Lost Man." *The New Yorker*, June 16, 1997.

"Progress Is Seen in Guatemala Case." *New York Times*, March 25, 1992.

Recer, Paul. "Examination of Ancient Skulls Prompts Theory about Human Settlement of the Americas." Associated Press, August 2, 2001.

Rensberger, Boyce. "Putting a New Face on Prehistory." *Washington Post*, April 15, 1997.

Wilford, John Noble. "New Answers to an Old Question: Who Got Here First?" *New York Times*, November 9, 1999.

BOOKS

Berlin, Ira. *Many Thousands Gone: The First Two Centuries of Slavery in North America.* Cambridge, Mass.: The Belknap Press of Harvard University Press.

Gross, Daniel R. *Discovering Anthropology.* Mountain View, Calif.: Mayfield Publishing Company.

Hening, William Waller. *The Statutes at Large; Being a Collection of all the Laws of Virginia.* New York: R. & W. & G. Bartow, 1823.

McCartney, Martha W. "A Study of the Africans and African Americans on Jamestown Island and at Green Spring, 1619–1803." National Parks Service, U.S. Department of the Interior, Williamsburg, Virginia, 2000.

Owsley, Douglas W., and Richard L. Jantz. *Skeletal Biology in the Great Plains: Migration, Warfare, Health, and Subsistence.* Washington, D.C.: Smithsonian Institution Press, 1994.

Owsley, Douglas W., and Jerome C. Rose. *Bioarcheology of the North Central United States: A Volume in the Central and Northern Plains Archeological Overview.* Fayetteville, Ark.: Arkansas Archeological Survey, 1997.

Stanford, Dennis. *The Walakpa Site, Alaska: Its Place in the Birnirk and Thule Cultures.* Washington, D.C.: Smithsonian Institution Press, 1976.

CORRESPONDENCE TO AUTHOR

Bass, Bill. Letter re: Doug Owsley's graduate school years. September 11, 2000.

Brown, Ryan. Fax re: Benton County officials meeting with Army Corps officials. November 5, 1999.

Chatters, Jim. E-mail re: Kennewick Man. May 8, 2001.

Dansie, Amy. Letter re: Spirit Cave mummy. June 10, 2001.

Hawkinson, Cleone. E-mail re: Trip with Doug to inventory Kennewick Man. July 11, 2001.

Lyon, Mike. Letter re: Childhood years in Lusk, Wyoming, with Doug Owsley. January 15, 2001.

Owsley, Douglas. Letter re: Larson Massacre dates. July 27, 2001.

———. Letter re: St. Mary's City. December 28, 2001.

Schneider, Alan. Fax re: Bruce Babbitt decision declaring Kennewick Man a Native American. July 17, 2001.

———. Letter re: Kennewick Man inventory. July 12, 2001.

Stanford, Dennis. E-mail re: "Anzick Chapter." July 16, 2002.

LAWS

Archaeological Resources Protection Act, 16 USCA 470aa.
Native American Graves Protection and Repatriation Act, 25 USCS 3001.

PRESS RELEASES

"BLM Makes Spirit Cave Man Determination," Bureau of Land Management, August 15, 2000.
"Interior Department Determines 'Kennewick Man' Remains to Go to Five Tribes," U.S. Department of Interior, September 25, 2000.

REFERENCE MATERIALS

The American Heritage Picture History of the Civil War. American Heritage Publishing Co., 1960.
Buikstra, Jane E., and Douglas H. Ubelaker. *Standards: for Data Collection from Human Skeletal Remains.* Arkansas Archeological Survey Research Series no. 44, 1994, Fayetteville, Ark.
Haglund, William D., and Marcella H. Sorg. *Forensic Taphonomy: The Postmortem Fate of Human Remains.* New York: CRC Press, 1997.
Hall, Kermit L. *The Oxford Companion to the Supreme Court of the United States.* New York: Oxford University Press, 1992.
Smithson, James, founder of the Smithsonian Institution. All material pertaining to him was obtained from the Smithsonian's on-line archives, available at www.si.edu/archives/documents/smithson.htm.
Ubelaker, Douglas H., and William M. Bass, Richard L. Jantz, and Fred H. Smith. *A Review of Human Origins.* 6th ed. 1990. (Dr. William Bass at the University of Tennessee, Knoxville, TN provided this to the author.)
Wilshin, Francis F. *Manassas (Bull Run).* Washington, D.C.: National Park Service Historical Handbook Series no. 15. 1957.

REPORTS

Babbitt, Bruce. Electronic copy of the original notice announcing that Kennewick Man is Native American. September 21, 2000.

Huckleberry, Gary, and Julie K. Stein. "Kennewick Man: Analysis of Sediments Associated with Human Remains Found at Columbia Park, Kennewick, WA." 1999.

McManamon, Francis P. "Kennewick Man: The Initial Scientific Examination, Description, and Analysis of the Kennewick Man Human Remains." 1999.

Owsley, Douglas. "Report of the Inventory of the Kennewick Skeleton." October 28–29, 1998.

Powell, Joseph F., and Jerome C. Rose. "Kennewick Man: Report on the Osteological Assessment of the 'Kennewick Man' Skeleton (CENWW.97.Kennewick)." 1999.

ACKNOWLEDGMENTS

Words of acknowledgment are inadequate to describe the contribution my wife Lydia made to this book. During the three years that it took me to research and write *No Bone Unturned*, two of our three children were born; we moved twice; we built a home; I graduated from law school; and I ran for U.S. Congress in Connecticut. Through it all, Lydia bore what often felt like unbearable pressure with cool poise. I often say that behind every good man is a better woman. In my case, the man pales in comparison. I dedicated this book to Lydia because no one sacrificed more to bring it to publication.

Our children Tennyson Ford, Clancy Nolan, and Maggie May, although they are too young to know it, also sacrificed a lot while I wrote.

I am also indebted to the Owsley family. A book of this nature required, no, *demanded*, that I spend an exceptional amount of time with Doug. Interviews, even scores of them, were not enough. Doug Owsley is probably the most private, introverted individual I have ever come across. He does not open up easily. Nor does he enjoy talking about himself or his accomplishments. He is humble to a fault. The biggest challenge I faced in writing this book was getting at his human side, the personality behind the scientist.

In order to expose his extraordinary life and human qualities, I had to shadow him at work, at home, and on the road. I slept in his house, ate at his dining room table, rode in his car, and studied in his office. Doug's wife Susie, and their daughters Hilary and Kim, allowed me to be a steady intruder and made me feel more like a member of the family than a visiting journalist.

I also owe special thanks to Doug himself. We were complete strangers when I walked into his life with a notepad and tape recorder in October 1999. He extended to me the most cherished thing a journalist can obtain: trust. He and I were, respectively, a scientist and a journalist at the outset. Now we are on each other's short list of closest friends. My "off the record" time with him is among the most prized experiences I've had as a journalist.

This book is steeped in law and science. I have a degree in law. I know little of science. Besides Doug, I relied on George Gill, Richard Jantz, Bill Bass, and Dennis Stanford to educate me. Besides supplying me with academic papers and scholarly books, they granted scores of interviews. They also took the extraordinary step of reviewing portions of the manuscript prior to publication to ensure that the science aspects were correct.

As the scientist featured in the book, Doug always reminded me that he would not be where he is today without the tutoring and influence of Gill, Jantz, Bass, and Stanford. As a journalist, I can say the same thing. Without them, this book would read much differently.

Doug's assistants at the Smithsonian made my reporting easier and pleasurable. In particular, Kari Sandness and Rebecca Kardash ("Doug's Angels"), constantly answered questions, retrieved files, and tracked down documents. They were also instrumental in fertilizing the seeds of curiosity in my young son Tennyson, who accompanied me on two of my research trips to the Museum of Natural History. They may never know the contribution they made to his young mind.

Smithsonian photographer Chip Clark, besides being the most phenomenal photographer I've ever witnessed in action, was diligent and patient in tracking down hundreds of photographs that were used for the research phase of this book, and scores more that appear in the text. His pictures bring this book to life. Thanks, Chip.

Pam Stone was particularly helpful to me in researching and writing the Waco chapters.

Other individuals brought a real, human side to this manuscript. The Blake brothers, Sam and Randy, granted interviews in which they discussed highly sensitive details about their brother and his experience in Guatemala. They also supplied key documents and background research on the political conditions in Guatemala.

The people of Lusk, Wyoming, opened their homes and their hearts to me when I visited there to research Doug's childhood. I am particularly indebted to Doug's Sunday school teacher Idy Bramlet, his Cub Scout den leader Audrey Pfister, and Doug's boyhood friend Mike Lyon.

Thanks to attorneys Paula Barran and Alan Schneider, reporting on the legal aspects of this book was more fun than work. With their assistance I reviewed thousands of pages of the case file. They also permitted me an insider's view to their strategy and the prosecution of their case, welcoming me into their law offices and enabling me to report details that typically escape journalists.

The court staff at the U.S. District Court also aided me in Portland, as did the following: the Benton County coroner's office; the Benton County prosecuting attorney's office; the University of Wyoming's anthropology department; the University of Tennessee's anthropology department; Stafford Laboratories in Boulder, Colorado; Friends of America's Past; and the University of Nevada Museum, particularly curator Amy Dansie.

Cleone Hawkinson does not show up a lot in the manuscript. But her research and resourcefulness were crucial to my reporting.

The park service personnel and historians at Jamestown, St. Mary's City, and Yorktown supplied historical reports and granted interviews that enabled me to better understand the significance of these colonial sites.

The FBI's cooperation for the Waco chapters was essential, particularly the insights and cooperation from Dr. Joseph DiZinno.

Some other individuals whose names do not appear in this story were nonetheless instrumental in helping me tell the story. University of Connecticut psychologist Dr. Marguerite Capone guided me in the composition of interview questions that enabled me to better get at Doug Owsley's personality traits. Matt Eyring provided the Spanish translation for the Guatemala chapters. And Jim Rollins served as an extra set of eyes to gather important science journal articles.

The folks at HarperCollins have provided me time to do that which I enjoy most: write. I constantly remind myself how privileged I am to have a publisher. I appreciate the trust from publisher Cathy Hemming. I'm grateful for the friendship and help from Kyran Cassidy and Joelle Yudin.

Finally, my editor, Mauro DiPreta, and my agent, Basil Kane, are the brain trust that I work most closely with in the development and production of my books. This one was no exception. Mauro is like the surgeon who makes painful incisions. Only his knife is a pencil. His cuts and critiques of my various drafts are the result of hours and hours of contemplation and concern for my writing. Every author should be so fortunate to have an editor who cares so much. Basil is the master agent, an untiring sounding board. Our conversations about the scope and direction of this book probably exceed one hundred. He is the quintessential writer's best friend.

INDEX